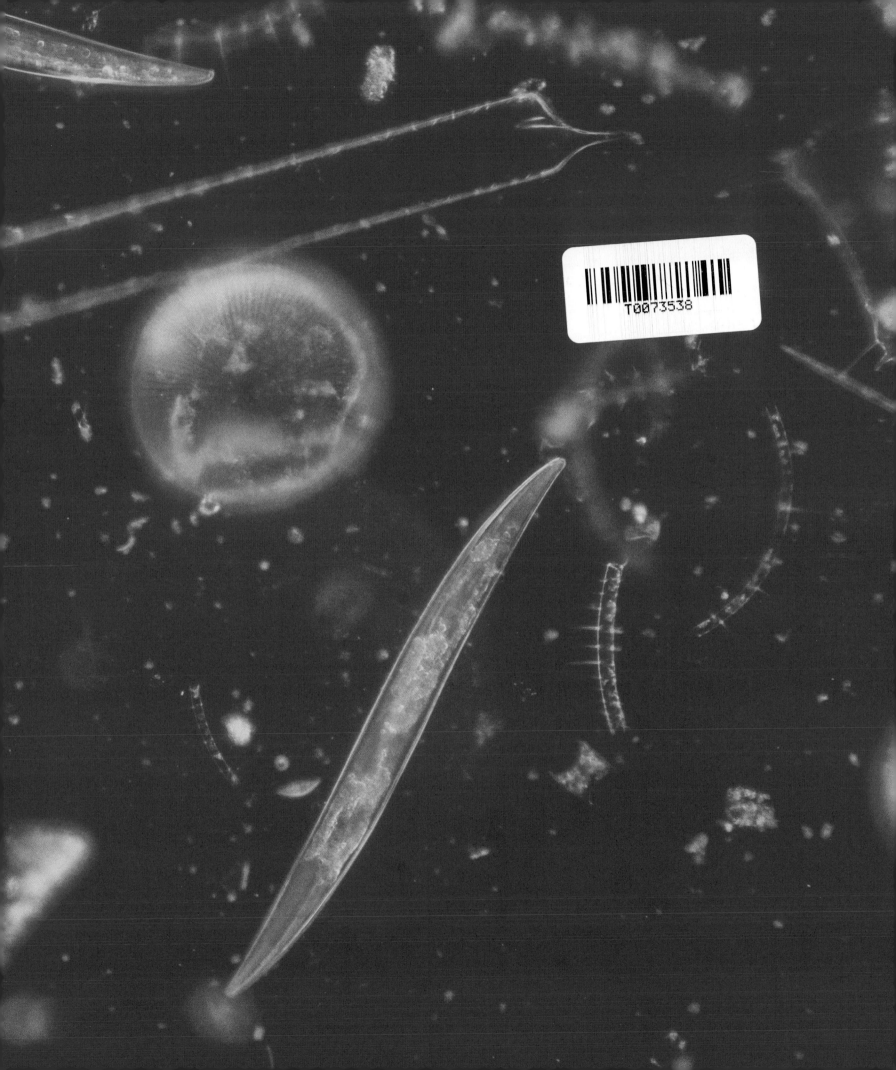

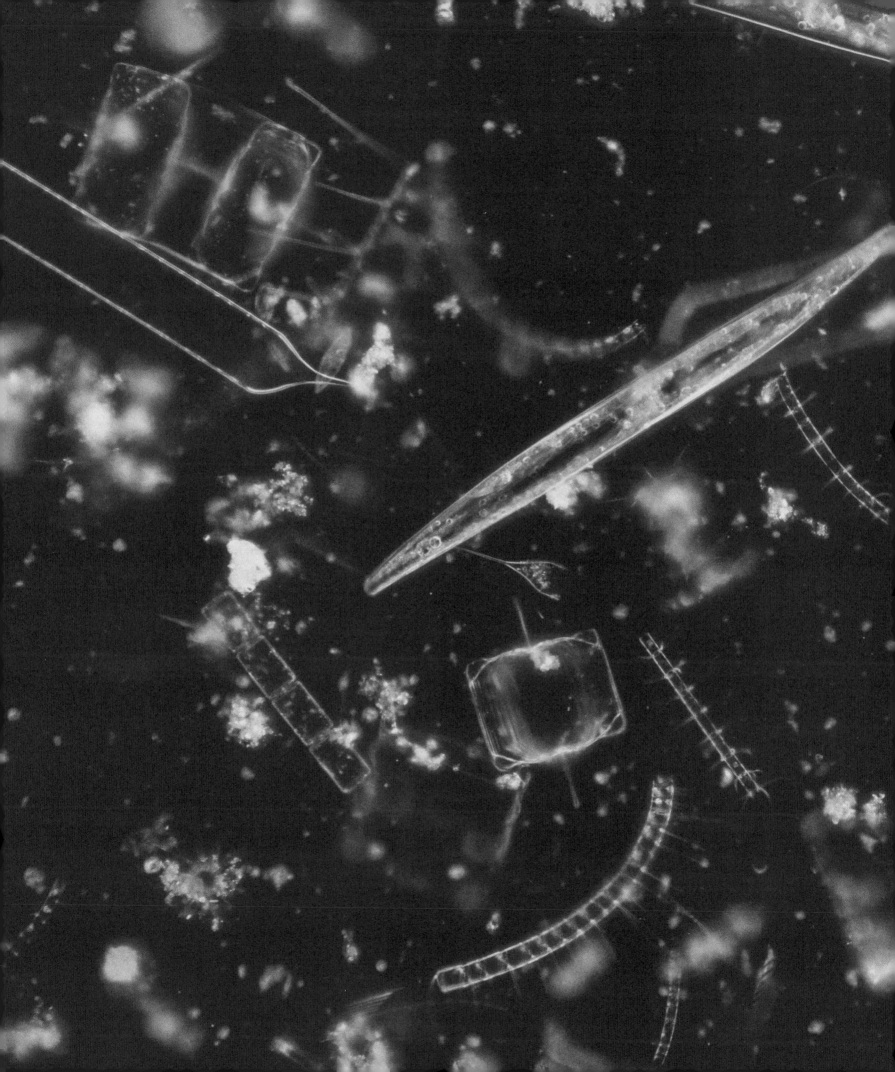

DEEP WATER

The University of Chicago Press, Chicago 60637
Text © 2023 by Welbeck Non-Fiction Limited
Design © 2023 by Welbeck Non-Fiction Limited

Published 2023
Printed in Dubai

32 31 30 29 28 27 26 25 24 23   1 2 3 4 5

ISBN-13: 978-0-226-82731-5 (cloth)
ISBN-13: 978-0-226-82733-9 (e-book)
DOI: https://doi.org/10.7208/chicago/9780226827339.001.0001

First published in 2023 by Welbeck
An imprint of Welbeck Non-Fiction Limited
part of Welbeck Publishing Group
Offices in: London – 20 Mortimer Street, London W1T 3JW &
Sydney – Level 17, 207 Kent St, Sydney NSW 2000 Australia
www.welbeckpublishing.com

Library of Congress Cataloging-in-Publication Data
Names: Black, Riley, author.
Title: Deep water : from the frilled shark to the dumbo octopus and from the continental shelf to the Mariana Trench / Riley Black.
Description: Chicago : The University of Chicago Press, 2023. | Includes index.
Identifiers: LCCN 2022055946 | ISBN 9780226827315 (cloth) | ISBN 9780226827339 (ebook)
Subjects: LCSH: Deep-sea biology. | Marine ecology.
Classification: LCC QH91.8.D44 B53 2023 | DDC 578.77/9--dc23/eng/20230126

LC record available at https://lccn.loc.gov/2022055946

Editorial: Isabel Wilkinson, Alison Moss
Design: Russell Knowles, James Pople
Picture Research: Steve Behan
Production: Marion Storz

DEEP WATER

FROM THE FRILLED SHARK
TO THE DUMBO OCTOPUS AND
FROM THE CONTINENTAL SHELF
TO THE MARIANA TRENCH

RILEY BLACK

The University of Chicago Press

Contents

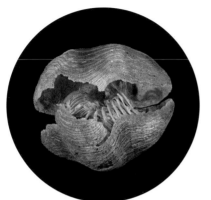

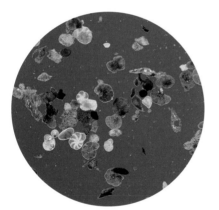

Timeline of Discovery

1817
Diel vertical migration discovered in *Daphnia*.

1836
Giant spider crab described by naturalists.

1843
Azoic hypothesis proposed by Edward Forbes.

1850
Fossil *Paleodictyon* described.

1857
Giant squid named by Japetus Steenstrup.

1872–76
Voyage of HMS *Challenger*.

1879
Giant isopods described by researchers.

1884
Frilled shark caught in Japan's Sagami Bay.

1884
Dumbo octopus described by naturalists.

1898
Goblin shark discovered off Japan.

1898–99
Vampire squid discovered off Africa.

1934
Record-breaking dive by the Bathysphere off Bermuda.

1938
Coelacanth caught and identified off South Africa.

1960
Trieste dives into the Challenger Deep.

1963
Feeding behaviour of the cookie-cutter shark discovered.

1965
First deep dive of DSV *Alvin*.

1976
The first known megamouth shark caught off Hawaii.

1977
Hydrothermal vents discovered off the Galapagos Islands.

1981
Riftia tube worms named by scientists.

1987
Whalefall ecosystem discovered off California.

1997
Second coelacanth species found in Indonesia.

2003
Tiburonia granrojo jellyfish named by scientists.

2003
Living *Paleodictyon* discovered on Mid-Atlantic Ridge.

2005
The first photographs taken of a giant squid in its natural habitat.

2006
Yeti crab *Kiwa* named from hydrothermal vent communities.

2014
Deepest dive by Cuvier's beaked whale recorded.

2018
Deep-sea stromatolites discovered in the Arabian Sea.

2018
Blubber discovered in fossil ichthyosaurs.

2019
"Tethered" sea squirt species discovered in Java Trench.

2020
Methane-eating bacteria found in the Pacific.

Introduction

WE LIVE ON AN ALIEN PLANET. THAT MIGHT SEEM LIKE A STRANGE SENTIMENT GIVEN THAT OUR SPECIES EVOLVED ON EARTH AND IT'S OUR ONLY HOME IN THE UNIVERSE. BUT EVEN THOUGH WE'VE SET FOOT ON THE MOON, SENT SATELLITES INTO SPACE, AND CONSIDERED HABITATION ON MARS, THE FACT OF THE MATTER IS THAT WE BARELY KNOW OUR HOME PLANET AT ALL.

When Looked at from space only about 30 per cent of the surface of our planet is land. The rest is water, and those waters run very deep. Beyond the fringes of continental shelf that jut from the edges of the world's land masses, the Earth's crust goes deeper and deeper, ultimately reaching cold and crushing depths at 10,984 metres (36,037 feet).

Our everyday world is not typical of life on Earth; if anything, it's the surface environment that's strange. Most of the planet is made up of deep-ocean habitats that are home to isopods larger than a football, devilish-looking squid that feed on plankton, sharks that migrate up and down the water column on a daily cycle to follow the microorganisms they feed on, and burbling vents surrounded by enormous worms that together might offer a hint at how life on our planet got its start. Most life on the planet is adapted to living in cold, dark conditions where the only lights are the ones created by the organisms that live there.

And yet we are inextricably connected to these mysterious and expansive habitats, from the boundary of the Twilight Zone at 1,000 metres (3,280 feet) to the deepest part of the Mariana Trench. Whales and other creatures dive deep to feed on organisms that dwell in darkness, returning those nutrients back near the surface in their bodily excretions. This effluvium provides sustenance for plankton, and many of those plankton eventually drift downwards to become part of the ever-shifting ocean floor. The deep sea is also part of our global carbon cycle that is changing because of human-created greenhouse gases. Carbon dioxide, methane, and other gases from the atmosphere are taken up by the upper layers of the ocean and used by photosynthetic plankton and creatures to make their shells. Vast numbers of these organisms sink deep to the bottom when they perish, piling up as sediments that eventually are compressed into rock and effectively bury the carbon, among other elements, they took up. Even if you never visit the deep waters below 200 metres (656 feet), you're still directly connected to it.

Opposite Coral reefs often form in the Photic Zone, or the uppermost layers of the seas, where sunlight penetrates.

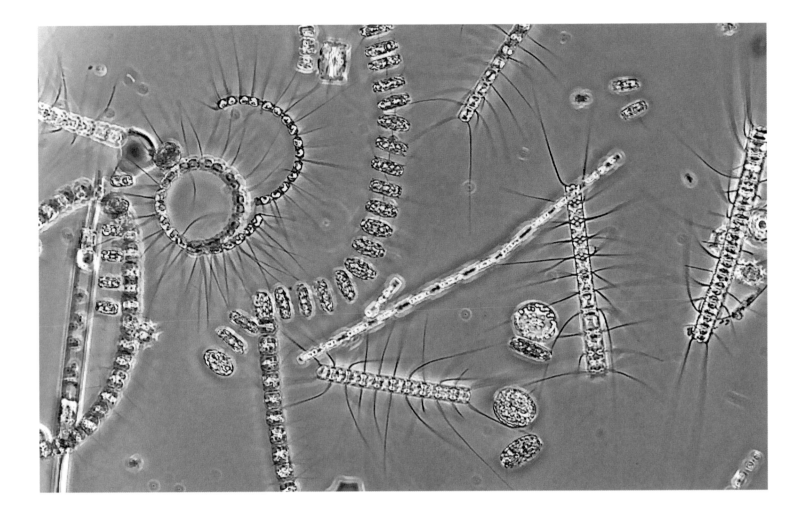

Our knowledge of the deep is still in its infancy. Despite all the innovations and technical breakthroughs in studying the oceans during the last 200 years, it's only very recently that we've been able to visit the deep ocean. Even when explorers visit the furthest reaches, like the Challenger Deep, they are only able to stay for a matter of minutes due to the immense pressure. It was only about 150 years ago that marine scientists realized that there is life to study below our familiar upper layers, and it's been less than 100 years since experts started to view the expansive abyssal plains – about half of the Earth's entire surface – as more than underwater deserts. Even when experts go searching, they are not always sure what they are actually looking at. It can take years for researchers to recognize that a striking jellyfish or fuzzy-looking crab is something no one has ever seen before. The fossil record of the deep sea is even more mysterious. Rocks that preserve life in deep habitats are very rare – partly because the deep sea is constantly creating new rock and recycling old rock – and so investigating this area is like trying to understand the history of life on our planet through a battered flipbook missing most of its pages.

But we are learning more. Every dive and every deep-sea creature that washes ashore tells us something new. What follows in this book is a selection of snapshots – creatures, concepts and environments – that have altered our understanding of Earth's natural history. Some were named centuries ago but have only recently been understood. Others are new discoveries that have shaken up what researchers previously assumed. All are part of a story that we are just starting to piece together. As you peruse each chapter, wondering over the behaviour of lanternfish or the biology of giant squid, remember this – all these organisms and environments are sharing our planet, right now, clothed in a deep darkness, their lives unfolding unseen in places that we can only briefly visit.

Opposite Oceans cover about 70 per cent of the Earth's surface and contain over 90 per cent of the world's water, making ours the "Blue Planet".

Above Phytoplankton form the basis of marine ecosystems. These photosynthesizing microbes are the foundation for food webs from the surface to the deep sea.

Zones of the Ocean

LOOK AT THE WORLD AROUND YOU. THE SURFACE OF THE EARTH IS COVERED WITH MOUNTAINS AND VALLEYS, PLAINS AND PLATEAUX, FROM CONTINENTAL DEPRESSIONS OVER 400 METRES (1,312 FEET) BELOW SEA LEVEL TO MOUNTAIN PEAKS THAT REACH OVER 8,800 METRES (28,870 FEET) IN THE AIR. AND YET ALL OF THIS, THE ENTIRETY OF EARTH'S TERRESTRIAL TOPOGRAPHY, COULD EASILY BE SUBSUMED BY THE SEA.

The Mariana Trench is deeper than Mount Everest is tall by thousands of metres/feet. As striking as they are, the world's continents only make up about 30 per cent of the Earth's surface. There is far more ocean – and deep ocean – than there is land, and the mysteries of these depths have led us to wonder "What's down there?" for century upon century. To humans, the deep sea may as well be another planet – a strange and even hostile place that requires special equipment to visit for even a few moments.

The ocean is constantly changing. Even the organisms that live within the deep sea, as we'll learn, do not always remain confined to the lower depths, but migrate up and down in the water column in search of food, light and mates. Nature does not fit neatly into boxes. Nevertheless, if we are to explore the deep sea, it's worth taking a moment to consider what *deep* truly means.

Above Advanced SCUBA divers can descend to 40 metres (131 feet).

Opposite above The Great Barrier Reef, off the east coast of Queensland in Australia, lies in the Photic Zone.

Opposite below Oceanographers generally recognize five major ocean zones, from the surface to the world's deepest trenches.

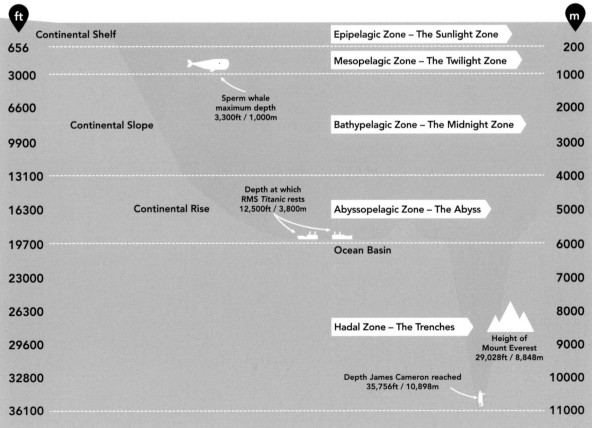

ft		m
	Continental Shelf	Epipelagic Zone – The Sunlight Zone	
656		200
		Mesopelagic Zone – The Twilight Zone	
3000		1000
	Sperm whale maximum depth 3,300ft / 1,000m	
6600		2000
	Continental Slope	Bathypelagic Zone – The Midnight Zone	
9900		3000
13100		4000
	Depth at which RMS Titanic rests 12,500ft / 3,800m	
16300	Continental Rise	Abyssopelagic Zone – The Abyss	5000
19700		6000
	Ocean Basin	
23000		7000
26300		8000
	Hadal Zone – The Trenches	Height of Mount Everest 29,028ft / 8,848m	
29600		9000
	Depth James Cameron reached 35,756ft / 10,898m	
32800		10000
36100		11000

Oceanographers and marine scientists often divide the oceans into five major zones, each with subdivisions within them. It's also worth noting that these divisions are primarily applied to parts of the oceans that are away from the coasts, beyond the continental shelves that skirt the edges of Earth's land masses. While both the coast and the open sea have a Photic Zone – the depths where light is still visible – the terminology for the deeper parts of the seas only come into play beyond the nearshore environments that experts know as the Neritic Zone.

So let's say that you're on a boat far from shore, on the open ocean beyond the reach of the continental shelf. Below you are five major ocean zones that correspond to increasing depths. Each has technical terms that may be used interchangeably or to emphasize a certain aspect of the depth, so to keep things simple we can call them by their common names.

The Sunlight Zone (Epipelagic Zone) comprises the top 200 metres (656 feet) of the ocean. A great deal of ocean biodiversity – from plankton to great toothy sharks – lives in this top layer. This top layer receives by far the most sunlight, which provides the raw energy photosynthesizing plankton need to make food and support so many of the oceans' food webs.

Sunlight begins to fade with depth. In fact, the Sunlight Zone is the shallowest of them all. At this zone's deepest extent, only about 1 per cent of the available sunlight is still visible. Little wonder that oceanographers consider this the border with the next ocean division – the Twilight Zone.

Technically called the Mesopelagic Zone, the Twilight Zone extends 200 to 1,000 metres (656–3,280 feet) below the ocean surface. Many famous deep-sea creatures live in this zone, from jellyfish that light up with bioluminescent colours to the still-mysterious giant squid. Comprising about 20 per cent of the entire

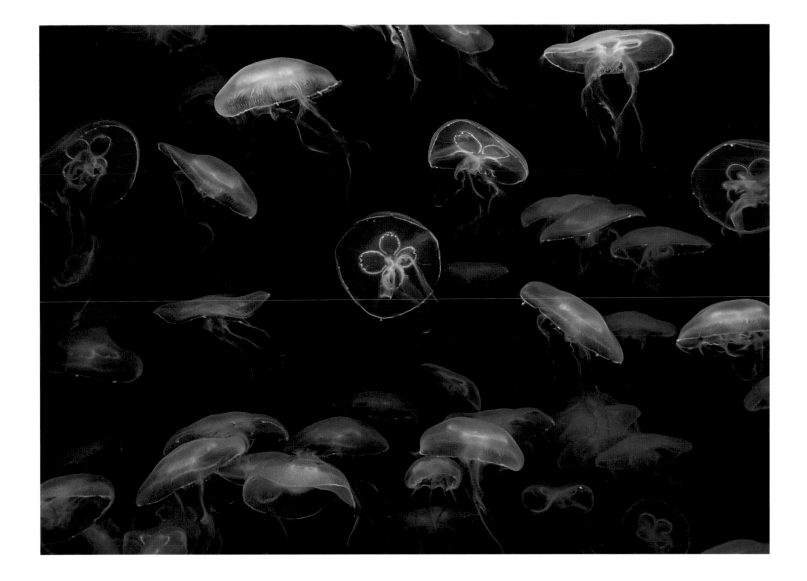

oceans' volume, this is the part of the seas that we often think of – a dark place where remote operated vehicles (ROV) shine their lights on otherworldly invertebrates and snaggle-toothed fish.

Yet the Twilight Zone is not particularly deep in the context of the oceans' entire depths. Below this zone, from 1,000 to 4,000 metres (3,280–13,123 feet) underneath the surface, is the Midnight Zone. Known as the Bathypelagic Zone to specialists, sunlight is not even a glimmer here. It's also incredibly cold. The average temperature remains about 4°C (39°F), with all the weight of the overlying water creating incredible pressure. And without sunlight, there is no photosynthesis. The organisms that live here must either feed themselves by consuming other organisms or by an alternative energy pathway called chemosynthesis – turning molecules such as carbon dioxide and methane into energy.

The ocean goes deeper still. Below the Midnight Zone, from 4,000 to 6,000 metres (13,123–19,685 feet) down, is the Abyssal

Zone (Abyssopelagic Zone). The comparatively few organisms that live here exist in total darkness. Temperatures are about 2°C (35°F) and the very bottom waters are often so depleted in oxygen that nothing can live there. Creatures that manage to survive in this hostile place tend to be so adapted to the cold, dark and pressure that they can't survive in the oceans' upper zones, and many feed on the detritus – or "marine snow" – that falls from above.

Below that, at depths greater than 6,000 metres (19,685 feet), lies the Hadal Zone (Hadalpelagic Zone) - the deepest reaches of the sea that can only be accessed by venturing into ocean trenches.

Opposite Remote operated vehicles – or ROVs – have been essential for exploring and mapping the seas.

Above Jellyfish can be found from the oceans' surface to over 3,700 metres (12,139 feet) down.

How Much of the Deep Sea is Unexplored?

IN THE DEEP SEA, THE DARK, THE CHILL AND THE PRESSURE PRESENT CHALLENGES FOR ALL FORMS OF LIFE – NOT LEAST TO OURSELVES. YET, AS OCEANOGRAPHER WILLIAM BEEBE ONCE WROTE, "ONE THING WE CANNOT ESCAPE – FOREVER AFTERWARD, THROUGHOUT ALL OUR LIFE, THE MEMORY OF THE MAGIC OF WATER AND ITS LIFE, OF THE HOME WHICH WAS ONCE OUR OWN – THIS WILL NEVER LEAVE US." WE ARE FASCINATED BY WHAT IS DOWN THERE, AND WHAT PRECIOUS LITTLE WE KNOW DRIVES OUR CURIOSITY STILL FURTHER.

According to the United States National Oceanic and Atmospheric Administration (NOAA), more than 80 per cent of the world's seas are uncharted. That estimate might seem shockingly high, but makes sense when you look below the surface. Even though humans have been travelling across the ocean for thousands of years, much of what we know comes from areas close to shore or relevant to our travels between land masses. We know even less about the oceans below the surface waters, into the dark Twilight Zone and below.

So far, NOAA estimates, only about 10 per cent of the world's seafloor has been mapped with modern sonar methods. And that's just telling us about the topography of the sea bottom, the seamounts and trenches and Abyssal Plains that lie far below the surface. We know even less about the organisms that live in these habitats. In 2022, researchers from the Natural History Museum in London announced that about 60 per cent of DNA sequences extracted from deep-ocean sediments could not be identified as animal, plant, bacteria, or something else. And that's just one study, based on what little matter experts have been able to bring back to the surface. Another estimate proposes that up to 91 per cent of the species that live in the oceans are undescribed and have never been seen.

Even though the oceans have been a critical part of our planet for billions of years, we have only just begun to visit and study them. It wasn't until 1930, when naturalists William Beebe and Otis Barton were lowered to a depth of 435 metres (1,427 feet), that anyone even saw the deep ocean with their own eyes (see pages 80–83). That is less than 100 years of exploration and innovation as we've tried to dive deep.

Naturally, submersibles like Beebe and Barton's Bathysphere are not our only sources of information about the deep ocean. Everything from deep-sea fish that wash up on the seashore to samples trawled up in nets to satellites capable of tracking ocean temperatures have informed our understanding of what's

Opposite An artist's impression of a bathysphere or deep-diving vessel. Oceanographers have envisioned a variety of ways of visiting the deep, from diving suits to submersible vehicles.

happening down below. Even so, the very nature of the deep sea makes it a difficult place to study.

Scientists sometimes comment that it's easier to send someone into space than it is for someone to visit the deep ocean. Pressure is a critical factor. We're used to the air pressure on land, and an astronaut going into space wouldn't feel any pressure at all. But the deeper you venture in the ocean, the greater the pressure. Even deep-sea animals can be limited by the increasing pressure – with ever-deeper species requiring specific adaptations to the harsh conditions – to the point that species living at the bottom

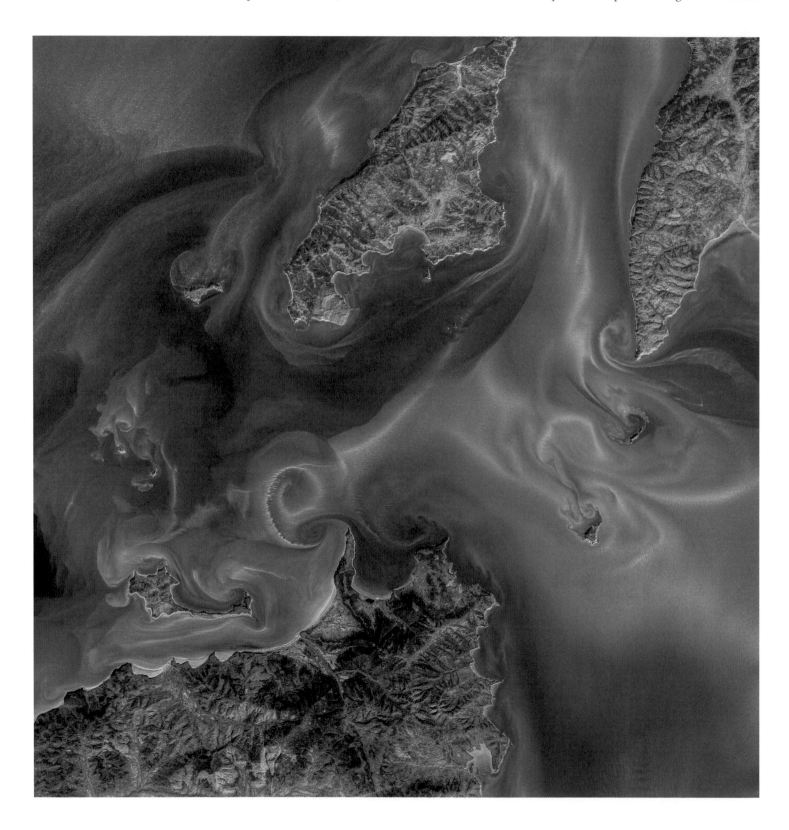

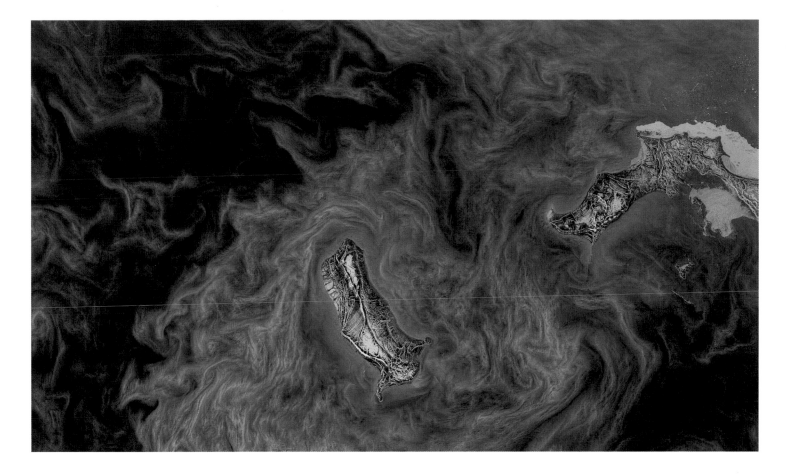

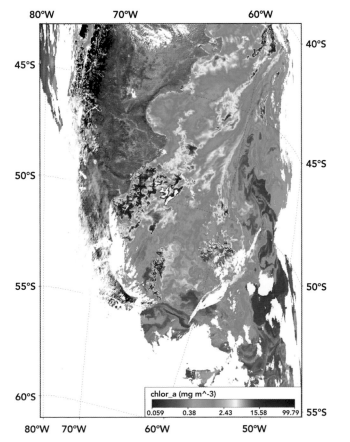

of the Mariana Trench (see pages 212–215) are under a thousand times more pressure than we experience at the surface. Designing remote and human-operated vehicles that can withstand the pressure and return safely to the surface is an incredibly challenging, delicate and expensive task.

But we must find out more about the deep ocean. Especially as humans have an ever-greater impact on the planet, we need to know what lifeforms are in the depths to better conserve and protect them. Even if shipping noise or human-caused climate change don't directly touch the deep seas, the depths are still connected to the surface through phenomena such as the daily migration of plankton and nutrient cycles, during which detritus and organic matter from the surface drifts down to feed life far below (see pages 20–25). Sometimes it seems that every time researchers venture into the deep sea, they find something new – an impression close to the mark, given all that remains to be discovered.

These pages Today's satellite images and data reveal a wealth of information about our seas and oceans, from tidal currents (opposite), to immense phytoplankton blooms (above), to temperatures (left).

Nutrient Cycling

THE OCEANIC DEPTHS ARE NOT DISCONNECTED FROM LIFE NEAR THE SURFACE. THE TWO ARE CLOSELY CONNECTED, NOT JUST BY THE FLOW OF CURRENTS OR THE MIGRATION OF PLANKTON BUT ALSO THROUGH THE BEHAVIOUR OF ANIMALS SUCH AS WHALES. THIS PHENOMENON IS CALLED NUTRIENT CYCLING, A BIOLOGICAL PUMP FOUNDED ON WHALE POOP.

Whales are permanently tied to the surface because of their need to breathe air, but many of the foods that cetaceans rely upon are found in the deep ocean. Sperm whales often feed on squid that live hundreds of metres/feet below the surface, and many baleen whales dive deep to sift krill out of the water column. These deep-water creatures are rich in nitrogen and iron, which end up becoming part of the whales' faeces as the invertebrates are digested.

Where the whales release all that nitrogen, phosphorus and iron matters for the health of the seas. So far as marine biologists know, whales tend to poop closer to the surface. Their faeces are often cloudy, or what one biologist has referred to as "oversteeped green tea", and they are released in a large plume through the water column. The enriched faeces disperse in the upper levels of the ocean, where sunlight reaches, and become critical resources for plankton.

The whale pump – as biologists sometimes call it – is the opposite of another important ocean phenomenon. Small particles of biological debris and organic matter are constantly drifting downward from the surface waters towards the bottom. This is called "marine snow," because it looks like flurries in the water column. This organic material often serves as food for deep-sea creatures, like the krill, and so the way whales feed concentrates and returns essential components back near the surface to maintain the ecosystem.

What the whales and other deep-diving marine mammals leave near the surface helps to grow a garden. Phytoplankton – tiny organisms in the ocean that can photosynthesize, like algae – benefit from the nitrogen and iron released by the whales. This plankton is the foundation for much of the oceans' ecosystems,

not to mention that more than half the oxygen in our atmosphere is generated by phytoplankton through photosynthesis.

But the way whales and other creatures nourish the seas has changed over the past several hundred years. Large-scale whaling decimated cetacean populations around the world, and despite conservation progress many populations have never recovered to historic levels. That means fewer whales to perpetuate the nutrient-cycle work, and fewer deep-sea compounds making it back up to the surface. Some studies, like one carried out based

Opposite Humpback whales at the surface.

Below Whales are an important part of nutrient cycling in the oceans, feeding deep and then leaving nutrient-rich waste products near the surface.

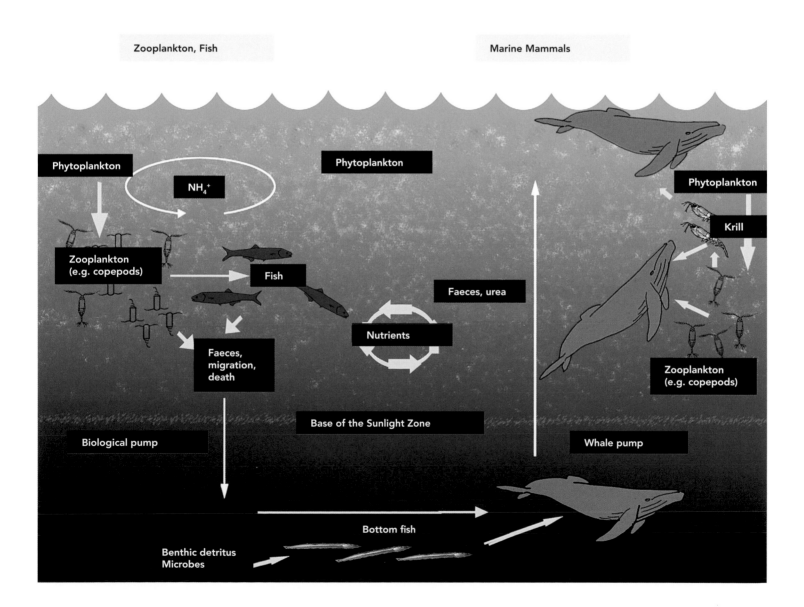

Above Humpback whales can dive over 200 metres
(656 feet) below the surface in search of food.

Opposite Centuries of intense whaling have dramatically
affected the nutrient cycling in the oceans.

Following pages Plankton are the most important organisms of the ocean. Everything
from armoured amoebas to tiny fish hatchlings form the basis of ocean food webs.

on the whale pump in the Gulf of Maine, off North America, estimate that whales and other marine mammals such as seals used to contribute three times more nitrogen to the surface before commercial culling. Another study found that nutrient cycling in the Southern Ocean, off Antarctica, is only at about 2 per cent of its former abundance.

Marine biologists have only recently come to understand the importance of large whales to the oceans' productivity and health. For decades, whales were thought to be large creatures that primarily consumed food in the ocean; we now understand them to be ecosystem engineers. That history has been going on for at least 40 million years, when whales began living permanently in the seas, which raises questions of how other creatures – like the marine reptiles of the Mesozoic era – might have affected the nutrients in ancient seas.

Naturally, life on land is connected to the seas. A baleen whale sifting krill in the ocean dark is important to our terrestrial realm. Whales help feed life in the surface waters, and in turn some creatures such as ospreys and other fish-eating birds take ocean fish and bring them inland – scattering those ocean nutrients on land. Those nutrients then become part of the terrestrial nutrient cycle that helps plants to grow, which feeds herbivores, and so on, with organic matter from the land being washed back out to sea in a never-ending cycle. In the past, researchers estimate, this interconnected system cycled over 150 million kilograms of phosphorus every year between the sea and land – an amount far greater than detected today. But there is good news. Through conservation and allowing sea life populations to climb, the oceans could return to a state closer to what they were like before whaling, harmful insecticides like DDT, and other ecological hazards. The oceans' past tells us what kind of future we can cultivate.

Bioluminescence

IF YOU STAND ON THE BEACH AT NIGHT, YOU MIGHT GET TREATED TO ONE OF NATURE'S MORE SPECTACULAR LIGHT SHOWS. WAVES SOMETIMES SPREAD TOWARDS THE SHORE WITH VIBRANT HUES OF NEON BLUE LIGHTING UP THE DARK SURFACE. THIS ISN'T AN OPTICAL ILLUSION: IT'S A COMMON FORM OF BIOLUMINESCENCE, A WIDESPREAD NATURAL PHENOMENON THAT TAKES PLACE FROM OUR BACK GARDENS TO THE DEEP SEA.

Reduced to its simplest expression, bioluminescence is light produced by a living thing. Many different organisms can bioluminesce, ranging from bacteria to fungi to fireflies. In short, light-emitting molecules called luciferins interact with enzymes called luciferases. But that's hardly all. These compounds are usually in bacteria that colonize the tissues of organisms, often in specialized light-emitting cells called photophores. From a firefly on a summer night to the glowing waves on the shore, it's this interplay that allows life to make light.

Sometimes bioluminescence is meant to advertise and attract, as in insects that flash lights to entice mates. But especially in the dark of the deep sea, where sunlight never reaches, bioluminescence can take on different roles – including camouflage. Even though lighting up bright might seem to make an organism more conspicuous, creatures that live in the Twilight Zone and below can emit light to startle or throw off their predators. Cookie-cutter sharks (see pages 98–99), for example, have a strong glow along their undersides – likely a way to make predators below them confuse the brightness of the shark's underbelly with the tone of the lighter waters above. Firefly squid use the same camouflage technique, and other creatures like lanternfish have photophores along the sides of their bodies to help disrupt the silhouette of their body.

Bioluminescent animals can pull some other flashy deep-sea tricks. Not only do the arms of the squid *Octopoteuthis deletron* emit light, but the cephalopod can jettison parts of an arm at will – and the glowing, twitching chunk can distract a predator while the squid makes its escape. A deep-sea shrimp called *Acanthephyra purpurea* uses a different technique to achieve similar ends. In addition to having photophores on the outside of its body, the crustacean can ooze a bright bioluminescent fluid, much like a glow-in-the-dark version of squid ink.

Of course, no glowing organism in the seas is quite so famous as anglerfish. These fish are not a singular species, but an entire order – called lophiiformes – that use specially evolved lures to help entice their prey. These striking fish have larger females that come to mind when we think of "anglerfish", while the males are tiny fish that only seek to attach to and parasitize the female (see

Opposite above Firefly squid are masters of ocean camouflage. The bright spots help break up their outline and confuse their shape.
Opposite below Firefly squid glow at the water's edge during mating.

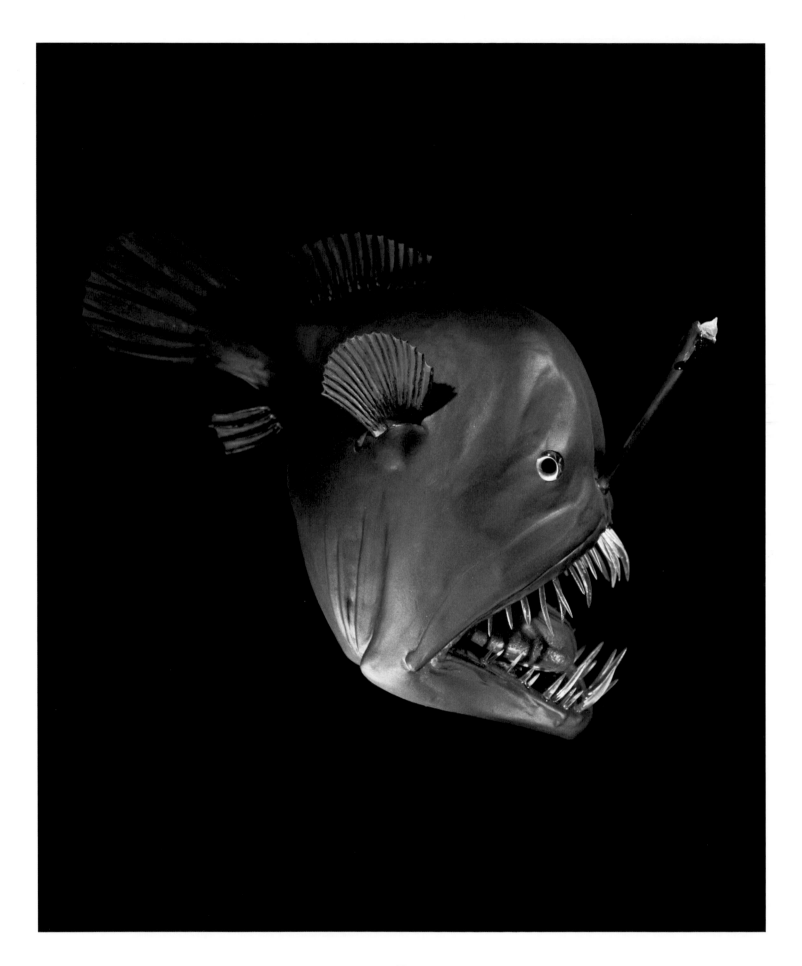

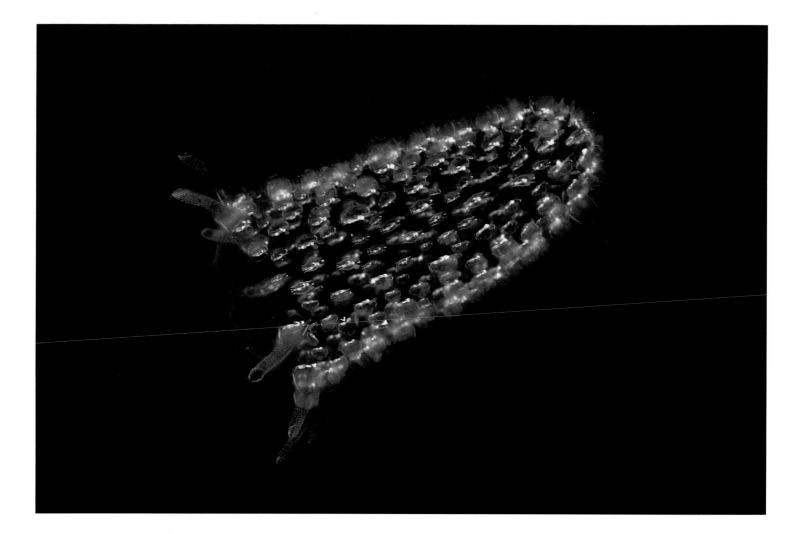

pages 132–135). And situated on the heads of these females, above a mouth full of pointed teeth, are lures that they can wiggle just so to entice prey. Among deep-sea species, those lures often glow.

In the case of the anglerfish, the bioluminescence comes from bacteria that contain the light-emitting compounds. The relationship between fish and bacteria is an example of symbiosis – when two different organisms become biologically reliant on each other in the same body – and the tie is so tight that some anglerfish bacteria can't survive outside this environment. How this relationship started is unknown, but researchers have detected that the bacteria inside the anglerfish lure are still evolving and losing genes.

Scientists have recorded an entire aquarium's-worth of glowing deep-sea species – squid, fish, corals, jellyfish and more. But as common as bioluminescence is, the sea isn't entirely aglow from top to bottom. In midwater, below the Sunlight Zone, about three quarters of known species can emit light. Along the sea bottom itself, the ratio of bioluminescent species drops to less than half. That makes sense for how the habitat shapes the lives of deep-sea species. Midwater is almost like space – expansive and seemingly borderless. Many species that live in midwater spend their entire lives in the ocean void. But the bottom is often murky and strewn with rocks and other obstacles that makes visual communication less effective, not to mention that many animals can hide by digging in rather than flashing.

There are probably many more deep-sea species that can glow and flash that are still to be discovered. Of the species we do know about, those that can give off light often only do so as a form of defence. By spending more time in the dark water, we may eventually be able to shed light on all the incredible ways these creatures make and use bioluminescence.

Opposite Anglerfish use bioluminescence to catch prey.

Above A bioluminescent tunicate. The light is created by the interaction between compounds called luciferins and enzymes called luciferases.

Following pages A lanternfish with bioluminescent fins and tail to disguise its shape.

Frilled Shark

IN 1884, AMERICAN ICHTHYOLOGIST SAMUEL GARMAN REPORTED ON A STRANGE FISH CAUGHT IN JAPAN'S SAGAMI BAY. THIS CREATURE, GARMAN WROTE, WAS "AN EXTRAORDINARY SHARK", AN EEL-LIKE SPECIES WITH ODD, MULTI-CUSPED TEETH THAT LOOKED LIKE LITTLE PITCHFORKS. ODDER STILL WERE THE SHARK'S GILLS, SIX SEEMINGLY "FRILLED" PAIRS INSTEAD OF THE FIVE SLITS MOST SHARKS HAVE.

Garman codified these strange features in the shark's name – *Chlamydoselachus anguineus*, the "eel-like shark with frills". Marine biologists weren't entirely sure what to make of the shark at first. The long, slender fish had a number of ancient traits that seemed to have more in common with fossil sharks than the likes of tiger and white sharks. Depending upon whose authority you relied, the frilled shark might be a surviving relative of early sharks such as the fossil *Cladoselache*, the spike-headed fossil shark *Xenacanthus*, or some other prehistoric group. In time, however, biologists realized that the frilled shark is a relative of today's cow sharks – or hexanchiformes – that are often found in the deep sea. Even though frilled shark fossils have been discovered, these are from the Pleistocene era or Ice Age, and indicate that the shark isn't quite the "living fossil" it's often been portrayed as.

Much like the distinctive goblin shark (see pages 88–91), the frilled shark can be found off the coast of most continents. Specimens have been spotted from New Zealand to the East Coast of the United States. The sharks seem to favour waters between 50 to 100 metres (164–328 feet) deep along the continental shelves, although they don't stay put in the deeper parts of their range. The sharks feed on fish, squid, smaller sharks, and other deep-water creatures that

Above Samuel Garman was the first ichthyologist to describe the frilled shark.

Opposite Frilled sharks often spend the daylight hours in the depths and come closer to the surface at night.

migrate up towards the surface at night. During the day, or when the upper water layers grow too warm, the sharks go deeper.

Studying the feeding habits of this unusual shark has been a difficult task, but marine biologists have been able to piece a few things together about the predator's behaviour. The frilled shark is not a fast swimmer and doesn't even have a particularly strong bite, but instead focuses on squishy prey like squid, or organisms that might be exhausted after spawning. Despite being a fluid medium, water is still viscous enough that frilled sharks can spread their pectoral fins to brace against the water and throw their spiky jaws forward in a rapid strike.

But the shark might have a sneakier method. The multiple rows of multi-pronged teeth lining the frilled shark's jaws are so densely packed that they appear lighter in colour, even in deep waters, acting as a kind of lure to small prey that might be attracted to or confused by the light. Once the prey gets close, the shark can open its jaws quickly enough to create suction and slurp the prey inside.

The sharks are often seen as solitary hunters, but, like many other deep-sea creatures, they do come together to socialize for at least one important event – mating. In 2008, researchers reported on more than 30 frilled sharks captured along the Mid-Atlantic Ridge over an undersea mountain. Such an aggregation of frilled

sharks had never been seen before. And encountering such an event truly was by chance. While not the deepest-dwelling sharks, frilled sharks still live deep enough that they're not directly affected by changing seasons above. Changes in light levels and temperatures from the surface don't regulate when they breed. The sharks reach sexual maturity when they reach about 1 metre (3 feet) or more in size, although how these sharks determine when to congregate is still a mystery.

Why is the deep ocean home to so many primitive-looking animals like the frilled shark, coelacanth (see pages 44–49) and vampire squid (see pages 66–69)? There's no single explanation, but rather several factors that affect our perceptions of evolution in the deep.

Despite the fact that most of our planet is ocean, and watery environments often have plenty of sediment to bury animal remains, fossils from deep-water environments are very rare. Sediments that turn to rock layers more commonly accumulate closer to shore, and become compressed over time. Those rock layers then shift with the continents and sometimes are brought back up to the surface, where they can be investigated by palaeontologists. Some gaps in the fossil record of deep-sea animals – like the 66 million years between fossil and living coelacanths – is created by an imperfect geological record. In the case of frilled sharks, most fossils are from ancient, shallow-water environments. When the sharks began inhabiting deeper waters, they were not preserved as readily.

The deep sea is also less susceptible to the perturbations that are common at or near the surface. There are no strong ocean currents, dramatic seasonal changes in light and temperature, or direct interference by humans. Deep-sea niches may last longer and be more stable, requiring less change for creatures suited to a particular way of life. Today's frilled shark is not unchanged from its ancestors, but its very characteristics and adaptations still suit a life in the dim and the dark.

Opposite Frilled sharks can often be identified by their distinctive, multi-cusped teeth. The bottom image shows a tooth row from the lower jaw.
Above While most species of living sharks have five gill slits, the frilled shark has six.

Biogenic Sediment

WHERE DID THE SEAFLOOR COME FROM? IT MIGHT BE EASY TO THINK THAT THE SEAFLOOR HAS ALWAYS EXISTED, AN UNDERSEA LANDSCAPE THAT REMAINS VIRTUALLY UNDISTURBED COMPARED TO OUR TERRESTRIAL REALM.

But the seafloor is constantly, imperceptibly changing. From undersea volcanoes to dust carried out to sea by the wind, the sea bottom comes together from a variety of different sources – including the organisms that live in the oceans. A great deal of ocean sediment is what researchers call biogenic ooze.

Even though a great deal of Earth's sediments are made up of older stone that has been broken down into tiny pieces, biogenic ooze is the remnant of life itself – a slurry of biological titbits that are more resistant to breaking down. Think of all the life

in the seas – not just the ocean giants or creatures visible to the naked eye, but the plankton and other microorganisms that are so numerous as to be beyond counting. It's the remnants of these organisms that make up much of the muck on the seafloor.

A portion of the world's biogenic ooze is made up of hard parts from large animals, or at least ones big enough to easily see. The beaks of squid, the scales of fish, the teeth of sharks, the exoskeletons of crustaceans and more can become part of the sea bottom as they are shed or discarded and drift down into the

abyss. But all of those larger organisms rely on a much broader and prolific array of tiny creatures to survive, many of which form their bodies or protect themselves with hard tissues.

Earth's seas are absolutely brimming with armoured life forms. Foraminiferans, for example, are essentially amoebas that grow shells around themselves and are part of the oceans' plankton. Coccolithophores are almost like an algal equivalent, plants that form hard discs and are often arranged in balls called coccolithospheres. And, as gardeners might know, the oceans are also full of diatoms – a single algae cell that is enclosed in a shell – that are sometimes sold as "diatomaceous earth" in garden shops. These organisms and more – such as tiny crustaceans called amphipods, or protozoans called radiolarians – live and die by the billions upon billions upon billions, and when they perish, their hardened bodies sink towards the bottom. They are part of a constant rain from above called marine snow, and, like snow, they accumulate in soft drifts as the seafloor is constantly remade.

This accumulation doesn't happen in the span of hours or days. Depending on how close to shore plankton or other hard parts begin to sink, the components of the biogenic ooze might travel kilometres/miles down before touching the bottom. The process can turn into a journey of more than a decade, acting as a record of what life was like in the upper zones of the ocean. In fact, geologists and other researchers interested in Earth systems often sample biogenic ooze to better understand how phenomena such as climate have shifted over time.

When organisms like foraminiferans and coccolithophores make their shells, they incorporate isotopes of oxygen from seawater into their hard parts. The nature of these oxygen isotopes is affected by local seawater, which is in turn influenced by how much fresh water is frozen into ice sheets and glaciers. During colder climates, when there is a greater amount of ice at the poles, planktonic

Opposite The white chalk cliffs on England's southeast coast were formed from calcareous oozes during the Cretaceous period.

Above Foraminiferans are so numerous that their hard "tests" can form sediments in the deep sea.

shell-builders incorporate O18 into their hard parts. When the climate warms and glacial meltwater changes the ocean chemistry, however, seawater has more O16, and so the ratio shifts. Based on this relationship, experts can look at samples from biogenic oozes – either fresh or ancient – and track how the global climate has shifted between warm and cool conditions through time.

Each of the hard-bodied organisms in the seas uses different minerals and molecules to create their tough outer coverings. Diatom tests are often made of silica (SiO_2), while foraminiferans and coccolith shells are made of calcium carbonate ($CaCO_3$). Biogenic ooze that's primarily made of foraminiferans or coccoliths is often called calcareous ooze; it can be compressed and transformed into chalk over time. The world-famous White Cliffs of Dover, for example, are calcareous oozes that formed at the bottom of the sea during the Cretaceous period – a term that itself means "chalk". Even though the surface of the sea might seem uniform or plain, these cliffs are a cross section that testifies to the ongoing accumulation that underlies our oceans.

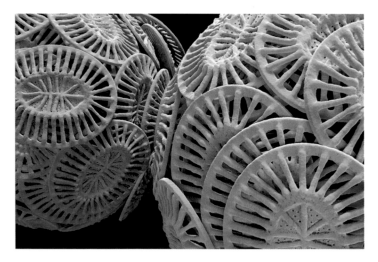

Top The ocean is full of invertebrates with hard shells, such as amphipods.

Above Coccolithospheres are made up of many individual coccoliths, or disc-shaped algae.

Opposite Many deep-sea sediments are principally made of diatoms that have drifted to the bottom.

Megamouth Shark

ON 15 NOVEMBER 1976, THE US NAVAL VESSEL *AFB-14* WAS OFF THE HAWAIIAN ISLAND OF OʻAHU WHEN SOMETHING BECAME TANGLED IN THE SHIP'S ANCHOR LINE. SOMETHING BIG. WHEN THE LINE WAS RAISED TO THE SURFACE, A SHARK 4.5 METRES (14 FEET 9 INCHES) LONG WAS CAUGHT IN IT – A SHARK THAT COULDN'T IMMEDIATELY BE CLASSIFIED. IT HAD A PASSING RESEMBLANCE TO SOME OTHER SIZEABLE OCEAN SHARKS, BUT INSTEAD OF LARGE, SERRATED TEETH, THE NEW SHARK HAD ROWS OF TINY, TRIANGULAR TEETH THAT BETTER FIT THE LIFESTYLE OF A FILTER-FEEDER. FORMALLY DESCRIBED FOR THE FIRST TIME IN 1983, THIS WAS *MEGACHASMA PELAGIOS* – THE MEGAMOUTH SHARK.

Researchers are still puzzled by this shark. It's one of several large, filter-feeding sharks alive today – like whale and basking sharks – but evolved its method of sieving small prey from the water entirely independently. Even stranger, the megamouth belongs to the same group of sharks as the famous great white and mako sharks, the lamniformes, and has fossil relatives dating back at least 34 million years. As far as marine biologists presently know, adult megamouth sharks can get to be about 5.5 metres (18 feet) long and live between the surface and depths of 1,000 metres (3,280 feet) in tropical and semi-tropical waters around the world.

Part of what makes this shark so difficult to understand is that much of what researchers have been able to observe comes from dead specimens rather than glimpses of the living animal. Examinations of dead megamouths have revealed that the inside of the shark's upper lip seems to be a bright whitish colour. At first, experts thought that this strip glowed in the deep – a bioluminescent lure for plankton to come closer. But later studies found that the strip doesn't produce light and instead is very good at reflecting it. It's possible that these reflectors help megamouths find each other in the deep, or maybe they play some role in feeding, or perhaps something else entirely, but it's difficult to tell when unobtrusively observing the sharks is so challenging. So far, only about 100 megamouth sharks have either been seen or caught.

Nevertheless, a few chance encounters have given biologists clues about how megamouths spend their time. While *Megachasma* can certainly be called a deep-sea shark, the filter-feeder doesn't spend all of its time below the Sunlight Zone. In 1990, researchers radio-tagged and released a megamouth shark and analyzed the data sent back. The shark never moved particularly fast, cruising along at 1 or 2 kilometres (about 1 mile) per hour, and it spent most of the daylight hours between 120 and 160 metres (393–525 feet). But at night, as plankton rose closer to the surface, the shark followed and stayed between 12 and 25 metres (39–82 feet) down, before returning to the depths in the morning.

While streamlined, the megamouth shark doesn't look quite so sleek and lithe as other shark species. Many sharks have stabilizing features near the base of the tail called caudal keels, for example, and the megamouth lacks these. But that makes sense for a slow-moving filter-feeder. Despite following plankton up through the water column at night, megamouths don't chase food – all they have to do is open their mouths, and the specialized rakers on their gills strain plankton and other small prey from the water. They live at such depths, and are so big, that adult megamouths are not usuallly in danger from predators – much like the giant oarfish (see pages 100–103). While other sharks have adapted to be warm-bodied, supercharged carnivores, the megamouth lives a slow-and-steady lifestyle reliant on abundant prey.

The sharks live such a relaxed lifestyle, in fact, that there isn't much genetic difference between megamouth sharks found in different oceans. This is often a major query for biologists – how many species or distinct populations of a species are there? However, in the case of the deep sea, the conditions far down are often so stable – and relatively similar around the world – that large animals can venture far and wide and be part of a global population rather than differentiated pockets. But there are some blank spots in the fossil record of megamouth sharks, so the next "new" *Megachasma* species is more likely to be found in the rock than in the seas.

Opposite The first megamouth shark was discovered when the fish became accidentally entangled in a US Navy ship's anchor.

Above The megamouth shark is a filter-feeder.

Following pages Living megamouths have rarely been photographed.

Coelacanths

THERE WAS A STRANGE, RARELY SEEN FISH THAT LIVED OFF THE COAST OF SOUTH AFRICA. PEOPLE THERE KNEW THE ODD CREATURE AS "GOMBESSA" OR "MAME", A FISH THAT SEEMED TO LIVE DEEP DOWN AND ONLY OCCASIONALLY MADE AN APPEARANCE IN FISHING NETS. BUT IT WOULDN'T BE UNTIL 1938 THAT WESTERN SCIENTISTS WOULD KNOW OF THE GOMBESSA, AN ANIMAL THEY THOUGHT LONG EXTINCT.

What unfolded on 23 December 1938 would soon turn the gombessa into a star. That morning, trawler captain Hendrik Goosen called the curator of the small museum in East London, South Africa – Marjorie Courtenay-Latimer. His nets had brought up a particularly strange fish and he had set it aside for Courtenay-Latimer in case she might want it for the museum. The crew took great care with the fish, trying to preserve it as best they could for their arrival at the harbour, despite the fact that the original dark blue of the fish's scales was fading to grey on their approach.

Courtenay-Latimer immediately knew that the fish was unlike anything in the museum, though she wasn't entirely sure what it was. With the local fish expert out of the office for the end-of-year holidays, she had a taxidermist preserve the fish as best they could. When James Leonard Brierley Smith eventually saw the taxidermized mystery, he recognized it as a fish that was thought to have been extinct since the time of *Tyrannosaurus Rex*. For scientists, the fish was proof that coelacanths still lived.

Coelacanths have a very long history. These bony fish first evolved about 410 million years ago, part of a diverse family called sarcopterygians. While they had interior skeletons made of bone, like many other fish of that time, their fins were markedly different. Instead of the fleshy fins of a shark, or fins made of a membrane stretched over fin rays, like most fish alive today,

sarcopterygians have many large, hand-like bones inside fleshy fins. This anatomical setup actually makes coelacanths more closely related to land-dwelling vertebrates – including our amphibious ancestors – than other fish.

Despite being prolific and sometimes reaching fantastic sizes – such as the Cretaceous coelacanths that grew to be as large as today's great white sharks – the fossil record of coelacanths almost entirely vanishes at the 66-million-year-old mark. (There have been some supposed coelacanth fossils dating to the past 66 million years, but these are rare and still questioned by experts

Opposite Marjorie Courtenay-Latimer's quick thinking and persistence
were essential to the scientific discovery of the coelacanth.

Top Coelacanths have an extensive fossil record, stretching back over 410 million years.

Above Preserved coelacanth (*Latimeria chalumnae*) found off the Comoro Islands, Indian Ocean.
The specimen has been bleached out by the preserving effects of formaldehyde.

45

given the difficulty in uncovering deep-sea fossils.) The discovery of a living coelacanth was like discovering a living descendent of *Triceratops*, a still-surviving remnant from the deep past.

Smith named the coelacanth caught off South Africa *Latimeria chalumnae*, a fish that could grow to 2 metres (6 feet 6 inches) in length, and eventually wrote a book about it called *Old Fourlegs*. But what no one knew was that the events of 1938 would be repeated. There was another living coelacanth species that would be discovered decades later.

On 18 September 1997, an American ichthyologist and his wife were browsing the stalls at a fish market in Indonesia while on honeymoon when they spotted a weird – but familiar – fish. It looked like the gombessa, just with a brown body colour. When the couple posted photos from their trip online, ichthyologists immediately began to speculate that this was a second coelacanth species. Additional specimens eventually confirmed that guess – there is a second living coelacanth species, *Latimeria menadoensis*. In fact, despite looking so similar to each other, the last time the two coelacanth species shared a common ancestor was over 30 million years ago. Genetic analyses have also indicated that coelacanths evolve slowly, despite how long ago they split into different species, and this may account for why fish that have been evolving independently of each other for so long retain so many similarities.

Modern coelacanths seem to live around the border of the Sunlight Zone and the Twilight Zone, around 90–150 metres (295–492 feet) below the surface during the day. They often rise at night, hunting for cephalopods and fish in waters about 55 metres (180 feet) deep, before returning to deeper water. Based upon where they live and how they spend their days, it seems that coelacanths prefer dim or dark waters which are neither warm nor cold – around 18–20°C (64–68°F). The temperature range seems to be important for the fish's physiology, yet coelacanths have a little trick to keep them from being too ecologically inflexible. If food seems scarce or the right conditions are hard to find, coelacanths can significantly slow their metabolisms to get by on less food and sink to lower depths to go into a sort of hibernation until conditions improve. If deep water has been a refuge for coelacanths since the last days of the dinosaurs, such a strategy might have allowed them to persist through the ages.

Left In life, coelacanths are a striking royal blue in colour.

Following pages Coelacanths belong to a group called sarcopterygians, or "lobe-finned" fish that have fins which are more anatomically similar to our limbs than to the ray fins of other fish.

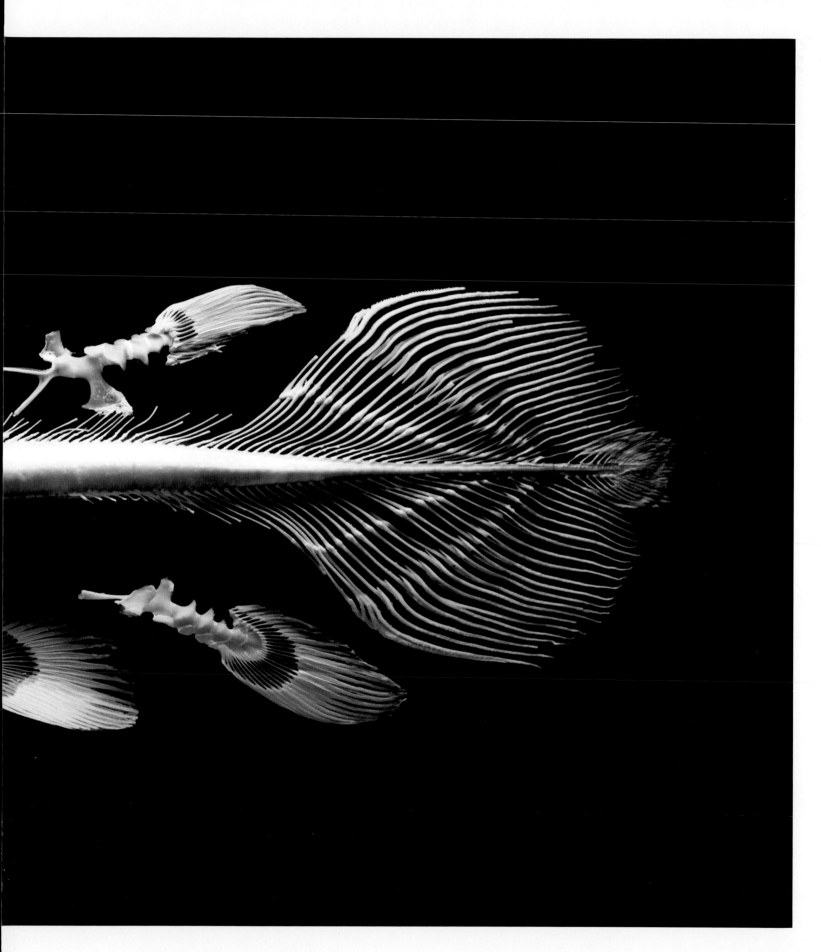

Azoic Hypothesis

ON THE SURFACE, A PAPER TITLED "REPORT ON THE MOLLUSCA AND RADIATA OF THE AEGEAN SEA" MIGHT NOT SEEM TO BE AN ESPECIALLY THRILLING DOCUMENT. PRESENTED BY NATURALIST EDWARD FORBES TO THE BRITISH ASSOCIATION FOR THE ADVANCEMENT OF SCIENCE IN 1843, THE DOCUMENT COVERED THE VARIOUS BIVALVES, CEPHALOPODS, STARFISH AND OTHER INVERTEBRATES FOUND IN THE POCKET OF OCEAN BETWEEN SOUTHERN EUROPE AND ASIA.

Such taxonomic lists were common in the age of colonialism, when various parts of the world were being mapped and documented. But Forbes didn't shy away from the theoretical. His report also contained a hypothesis that would frame people's ideas about the seas for over two decades. The deep sea, Forbes suggested, was almost devoid of life.

Forbes based his idea on his experiences aboard HMS *Beacon* in 1832. Acting as the ship's naturalist, Forbes set about dredging the sea at various depths and locations within the Aegean Sea as the *Beacon* made its survey. His goal was to better understand where ocean species were found, both geographically and at depth – or what biologists call the distribution of species. Forbes noticed something curious. The deeper he dredged, the variety of species coming up seemed to shrink – not to mention the size of the creatures themselves. The deeper the water, it seemed, the more sparse life became.

In Forbes's view, the depths of the ocean seemed to be organized into a different set of zones to those we're familiar with now. The uppermost parts of the Aegean were full of brown kelp, fish, oceanic invertebrates and other forms of life. Forbes' next layer was the corraline zone, where plants became scarce due to diminished sunlight and corals were more common. Below that, Forbes thought, were deep-sea corals that acted as a last holdout for the few species that could survive the darkness and pressure. And below that? Seemingly nothing. Even though Forbes was not the first to suggest the idea, he had found what he thought was the direct evidence of how sea life faded out with each successive metre.

Forbes was not able to sample the deepest zones of the sea, his being a time long before submersibles or any appreciation of how truly deep the world's oceans are, but, based upon his samples and a little mathematical wrangling, Forbes proposed that depths below 550 metres (1,804 feet) would contain no life at all. The idea became known as the azoic hypothesis, azoic meaning "without animals".

The azoic hypothesis – sometimes called the abyssus theory – could have been a blip on the scientific radar. At the time,

our understanding of the deep sea was incredibly minimal, not least because researchers could only get a surface view (many nineteenth-century books present illustrations of sea life stranded on the beach, because knowledge of what the animals looked like underwater, in their natural habitat, was extremely limited). But other experts helped to bolster Forbes's notion. Naturalists were beginning to understand that pressure in the ocean increased with greater depth. In 1867, for example, geologist David Page noted that water became ever more compressed the deeper the ocean got. "At this rate of compression," Page wrote, "we know that at great depths animal and vegetable life as known to us cannot possibly exist – the extreme depressions of the seas being thus … barren and lifeless solitudes."

Page's confidence was misplaced. Some creatures that were already familiar to naturalists of the time, like the giant squid, lived at greater depths than Forbes proposed. Experts just didn't know how deep because it would be over a century and a half before anyone would see a healthy giant squid in its natural habitat. But even ship- and shore-based surveys were starting to turn up evidence that the seas didn't follow Forbes's expectations. Around the time of Page's considerations of pressure, Norwegian biologist Michael Sars reported more than 400 different species found at depths below 800 metres (2,624 feet) – far below Forbes's cutoff. Scottish naturalist Charles Wyville Thomson

found life even deeper down, dredging up evidence of organisms from below 4,000 metres (13,123 feet). If that wasn't enough, the voyage of HMS *Challenger* between 1872 and 1876 would come back with evidence of life from some of the deepest parts of the ocean, more than 8,000 metres (26,246 feet) down.

Naturally, Forbes was working with what he had. The Aegean is not the most biodiverse area of the world's seas, and the dredges that he used were relatively simple. Then again, decades before his pivotal paper, naturalists had been finding invertebrates such as worms and starfish from lower depths than those that Forbes suggested were lifeless. He probably just didn't know about it because not all scientific results at the time would become widely known. But, as historians of science have noted, there was a benefit to Forbes's conjecture. It got people to go out and look, to double-check and explore, and those explorations eventually revealed an abundance of life that would have likely thrilled such a pivotal figure in the history of oceanography.

Above left Edward Forbes proposed the azoic hypothesis in 1843, based on observations of creatures in the Aegean Sea.
Above right Charles Wyville Thomson was among the experts who would eventually disprove the azoic hypothesis.

Cambrian Creatures

FOR MOST OF EARTH'S HISTORY, THE OCEANS MIGHT HAVE SEEMED EMPTY. BACTERIAL LIFE AND ORGANISMS LIKE CYANOBACTERIA – WHICH HELPED OXYGENATE EARTH'S ATMOSPHERE – ABOUNDED, BUT THERE WERE NO MULTICELLULAR ORGANISMS OR ANIMALS. LIFE EVOLVED BY ABOUT 3.22 BILLION YEARS AGO, BUT THERE WAS NOTHING MORE THAN SINGLE-CELLED ORGANISMS FOR OVER A BILLION YEARS AFTER THAT – PALAEONTOLOGISTS EVEN CALL THE TIME PERIOD BETWEEN 1.8 BILLION YEARS AGO AND 800 MILLION YEARS AGO "THE BORING BILLION", AS LIFE SEEMED TO GO THROUGH A LONG, GRINDING LULL.

But by 541 million years ago, something astounding began to happen. Not only had early animals evolved, but they also began to flourish into an incredible variety of forms. This was the "Cambrian Explosion", and it included creatures of the deep sea.

The most famous fossil site documenting the Cambrian Explosion is British Columbia's Burgess Shale. It's taken decades for palaeontologists to comprehend what they're looking at when they study these fossils, with some creatures mistakenly restored upside down as part of the learning curve. There were spiky worms, animals that looked like pincushions, predators with compound eyes and grasping appendages set in front of a shutter-like mouth, all while our own predecessors were pencil-long, worm-like swimmers that didn't even have a backbone yet. But the Burgess Shale represents a shallow-water reef out on the continental shelf. Palaeontologists have had to look elsewhere to find out what was going on in the deep sea.

Kangaroo Island, off the south coast of Australia, is another hotspot for Cambrian fossils. The Emu Bay Shale contains fossils of many soft-bodied animals preserved in fine sediment. The creatures here come from deeper waters, estimated to be several hundred metres/feet below the surface, and the rocks formed at greater depth than the Burgess Shale. These fossils are also geologically younger, about 515 million years old, and include different species than those found elsewhere.

Among the finds of the Emu Bay Shale are eyes from creatures called anomalocaridids. They were often the predators of their time and were related to early arthropods. Encased in tough exoskeletons, these animals could grow to more than 1 metre (3 feet) long, had mouths like camera shutters, grasping arms in

Opposite The famous Burgess Shale is exposed in the mountains of British Columbia.

front of those mouths, and moved about by flapping wing-like fins along their sides. In fact, palaeontologists have speculated that such creatures might have helped set off the Cambrian Explosion. Anomalocaridids had compound eyes that were much better at seeing prey, which required that other species evolve new defences.

It's a wonder that the Emu Bay Shale fossils were preserved at all. The Cambrian was a world of invertebrates and soft-bodied creatures. Bone, the tissue that makes up most of our skeletons, had not evolved and so the hardest materials were structures like the keratin covering the bodies of ancient Australia's invertebrates. The corpses of these animals had to be buried rapidly, and even then experts might only find a partial fossil. At Emu Bay, though, palaeontologists have found anomalocaridid eyes that are so well preserved that individual lenses of the predators' compound eyes can be seen. Experts have found more than 30 of these eyes so far, some of which are 4 centimetres (1½ inches) in diameter: together, these specimens help to outline the lives of these unusual arthropods.

Fossil eyes from one species in the Emu Bay Shale, temporarily called *"Anomalocaris" briggsi* until it is officially described, are very large compared to those of its relatives. The eyes also seem to have an "acute zone" of larger lenses in the centre, where the eye's resolution is enhanced. Together, these facets of the animal's anatomy hint that *"Anomalocaris" briggsi* could see in relatively dim waters like those of the lower Sunlight and upper Twilight Zones of today's seas.

Fearsome as *"Anomalocaris" briggsi* might have looked, though, experts think that this animal fed on some of the ancient oceans' smallest inhabitants. Unlike some other anomalocaridids, the eyes of this particular species were not on stalks but set into the exoskeleton of the head. The arms of the animal were better suited to filtering plankton out of the water column than grasping worms or other small prey, as other *Anomalocaris* species did. Palaeontologists think that *"Anomalocaris" briggsi* looked upwards to detect plankton glinting in what little light penetrated the deep ocean waters, and sieved its meals from such planktonic clouds.

Animals like *"Anomalocaris" briggsi* were new on Earth. There had never been large filter-feeders before. But the animal helped to pioneer a niche that other organisms would evolve to fill time and time again, beginning interactions that can still be found in our modern seas, not unlike today's vampire squid or other invertebrates that pluck their meals from the water column.

Opposite Exceptional Cambrian fossils have been found in the Burgess Shale.

Above *Anomalocaris* and its relatives were among the largest animals of the Cambrian.

Following pages Life in the Cambrian ranged from sponges and worms to creatures with strange anatomies, such as *Anomalocaris*.

Giant Spider Crab

NO MAJOR AQUARIUM SEEMS COMPLETE WITHOUT GIANT SPIDER CRABS. KNOWN TO RESEARCHERS AS *MACROCHEIRA KAEMPFERI*, THESE ORANGE-AND-YELLOW CRABS LOOK OTHERWORLDLY. THEIR BODIES ARE NOT PARTICULARLY LARGE, ABOUT THE SIZE OF A WATERMELON, BUT THEIR FOUR LEGS AND TWO ARMS HAVE AN INCREDIBLY IMPRESSIVE SPAN – UP TO 3.7 METRES (12 FEET 2 INCHES) FROM CLAW TO CLAW. THE CRABS ARE ALSO QUITE HEAVY, UP TO 19 KILOGRAMS (41 POUNDS). AS EVERYDAY AS THEY MIGHT SEEM, THEY ARE TRULY EXCEPTIONAL INVERTEBRATES.

Known to the Japanese as *taka-ashi-gani*, giant spider crabs did not get their scientific name until 1836. In that year, Dutch zoologist Coenraad Jacob Temminck described the huge invertebrate – the species name *"kaempferi"* referring to German doctor Engelbert Kaempfer, purportedly the first European to see this creature. The crabs principally live around the Japanese archipelago at depths between 20 and 600 metres (65–1,968 feet), wandering around the soft sea bottom in search of food.

These crabs are huge – the largest crustaceans known – but they're still on the shrimpy side compared to science-fiction and horror-movie visions of enormous crabs. There's a good reason for that. Crustaceans wear their skeletons, made of a tough but relatively flexible material called chitin, on the outside. All of the crabs' internal organs – the nervous system, vital organs, musculature – is enclosed in this armour. That means that as crabs get larger and larger, they have more internal volume to decreasing surface area. There's a limit to how large they can be without becoming so heavy as to exceed the stress limits of their own exoskeletons. While it's possible that there could be as-yet-undiscovered crabs that are larger, the fact that the giant spider crab's large size comes from a relatively compact body with very long arms makes sense given the mechanical constraints of how these invertebrates build their bodies.

Naturally, living large comes with a mix of advantages and disadvantages. While larval *Macrocheira* start off life very small, the adult crabs are large enough that many potential predators are too small to eat them. A bumpy and spiky carapace helps offer another layer of deterrence. Yet that's hardly all. Giant spider

crabs, like some other crab species, decorate themselves with other organisms to help them blend in with the seafloor. There's no real strategy in what the crab selects. Almost anything will do – sponges or anemones or tube-dwelling worms, it's all the same to the crab. After plucking up the animal from the sea bottom, the crab fiddles with it to orient how the other organism will best stick and places the animal on top of its shell, where it becomes adhered. Some of these animals even begin to colonize and live on top of the crabs over time, a form of living camouflage that helps the crabs avoid detection as octopus swim by.

The crabs have become a rarer sight than they used to be in Japan's waters. That's because of humans. Giant spider crabs are sometimes eaten, and historic catches declined precipitously in the late twentieth century. Even the crabs that are hauled up today tend to be much smaller than in the past, with legs about 1 metre (3 feet) wide instead of nearly 4 metres (13 feet). The crabs were being taken at a faster rate than they could reproduce, and still require a great deal of help to recover.

The way the life cycle of the giant spider crab plays out is part of the challenge that conservationists face. Like many other deep-sea animals, giant spider crabs start off life very small – really part of the oceans' plankton. Rather than producing relatively few, well-developed offspring, giant spider crabs go for quantity; gravid female crabs can lay over a million eggs during a mating season.

From there, the crabs start to go through some quick changes. In less than two weeks, whatever eggs haven't been eaten by filter-feeders begin to hatch. About 15 minutes later, the crabs start to go through a series of growth phases, each stretching out longer than the previous ones. These tiny, developing crabs drift along near the surface for weeks before obtaining their familiar, mature shape, though it can take a decade before the crabs get anywhere close to their record sizes. Giant spider crabs are thought to live as long as 100 years, perhaps explaining why overfished populations tend to be so small.

Opposite Coenraad Jacob Temminck formally described the giant spider crab in 1836 in honour of colleague Engelbert Kaempfer.

Above Giant spider crabs can take decades to grow to their maximum size.

Following pages The largest spider crabs can grow to almost 4 metres (13 feet) across their spindly legs.

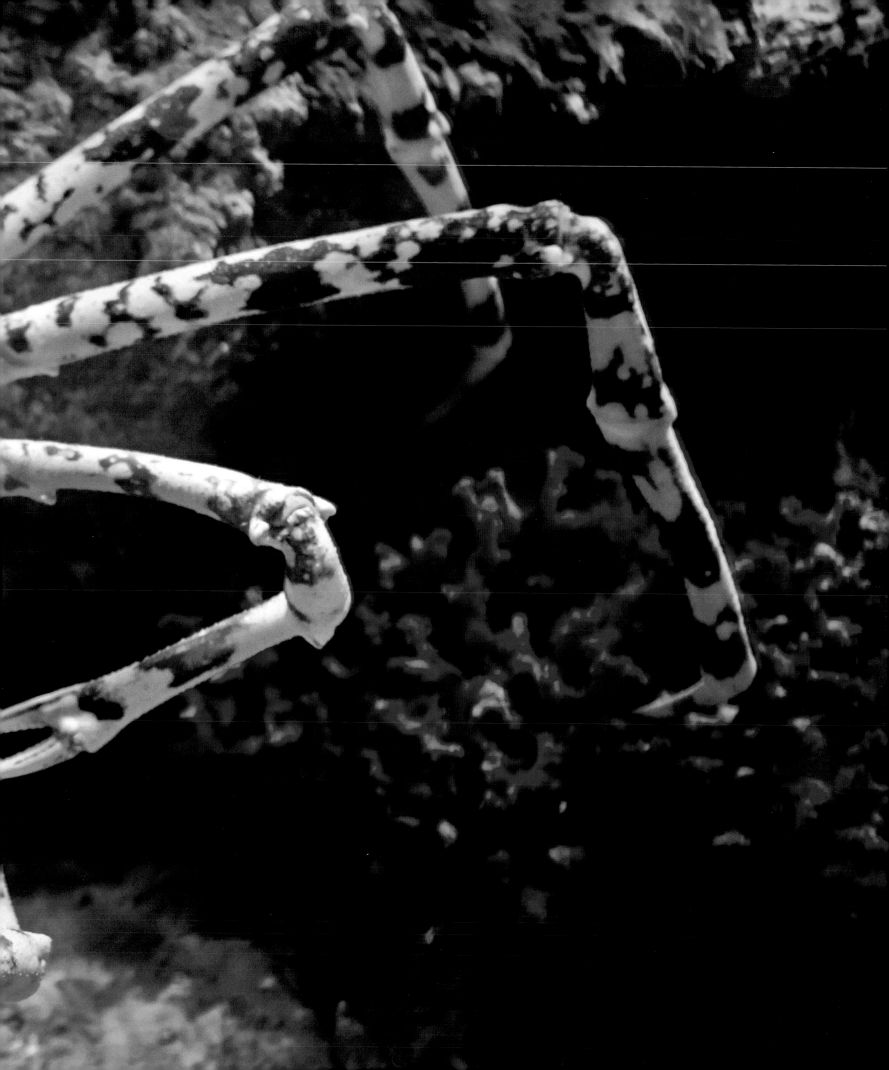

Ophthalmosaurus

TENS OF MILLIONS OF YEARS BEFORE DOLPHINS AND SEALS WOULD DIVE DEEP TO SNACK ON THE FISH AND SQUID THAT LIVE IN THE OCEAN DARKNESS, THERE WAS A REPTILE THAT MAY HAVE DONE THE SAME. NAMED *OPHTHALMOSAURUS ICENICUS*, THIS JURASSIC CREATURE WAS AN ICHTHYOSAUR – OR "FISH LIZARD" – THAT FLICKED THROUGH THE SEAS AT THE SAME TIME DINOSAURS DOMINATED THE LAND. AND BASED ON THE ENORMOUS EYES OF THIS REPTILE, SOME PALAEONTOLOGISTS SUSPECT THAT *OPHTHALMOSAURUS* DIVED VERY DEEP.

Ichthyosaurs were not dinosaurs; they were descended from a different branch of the reptile family tree. The very first ichthyosaurs evolved from land-dwelling ancestors over 236 million years ago. They were relatively small, less than 2 metres (6 feet 6 inches) in length, and looked like sinuous lizards with pointed snouts. But within 3 million years, those small swimmers had begun to evolve into a diverse array of ocean-dwelling reptiles. Some ate small prey while some hunted reptiles, including other ichthyosaurs. Some remained small while others grew to sizes comparable to modern sperm whales. Even as various forms of marine reptiles evolved and flourished, the ichthyosaurs were the greatest success story of the Mesozoic seas.

Ophthalmosaurus lived about 160 million years ago, in the later part of the Jurassic period, in seas that covered what is now Western Europe. These ichthyosaurs were much more streamlined than their early ancestors, and about the size of a modern dolphin. In fact, ichthyosaurs like *Ophthalmosaurus*

have often been used as an example of convergent evolution – to describe species that independently evolved the same shape, behaviour, or other similarity despite being distantly related. Both dolphins and some ichthyosaurs share similar body shapes, as both groups were air-breathing vertebrates that pursued small, quick prey in the water. Ichthyosaurs even had blubber to help keep them warm, much like whales of all sizes today. Even though there are some important differences between the oceans of the Jurassic period and today, *Ophthalmosaurus* and dolphins nevertheless fill the same sort of niche.

One of the most striking features of many ichthyosaur fossils is that they are preserved with the delicate bones inside the eye, called scleral rings. These bones can be informative clues about a creature's habits. How well an eye is able to see is dependent on various conditions – such as how large it is in absolute terms,

Opposite *Ophthalmosaurus* was a marine reptile that evolved a shark-like body to swim efficiently through the seas.

Above and below Exquisitely preserved fossils have provided palaeontologists with a detailed outline of what this reptile looked like, including the large eye socket.

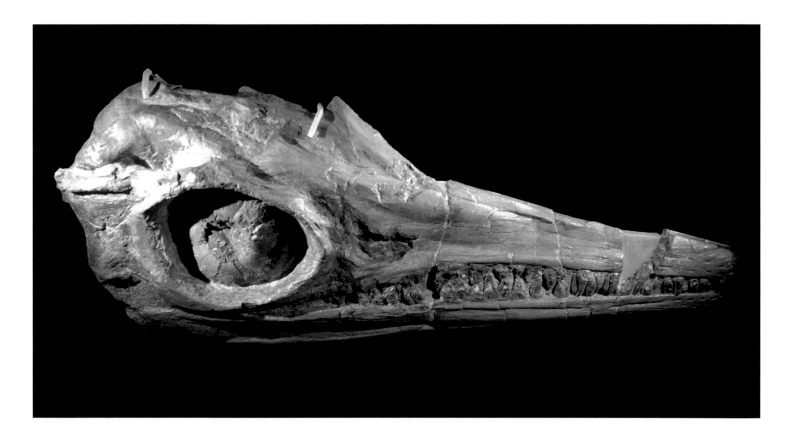

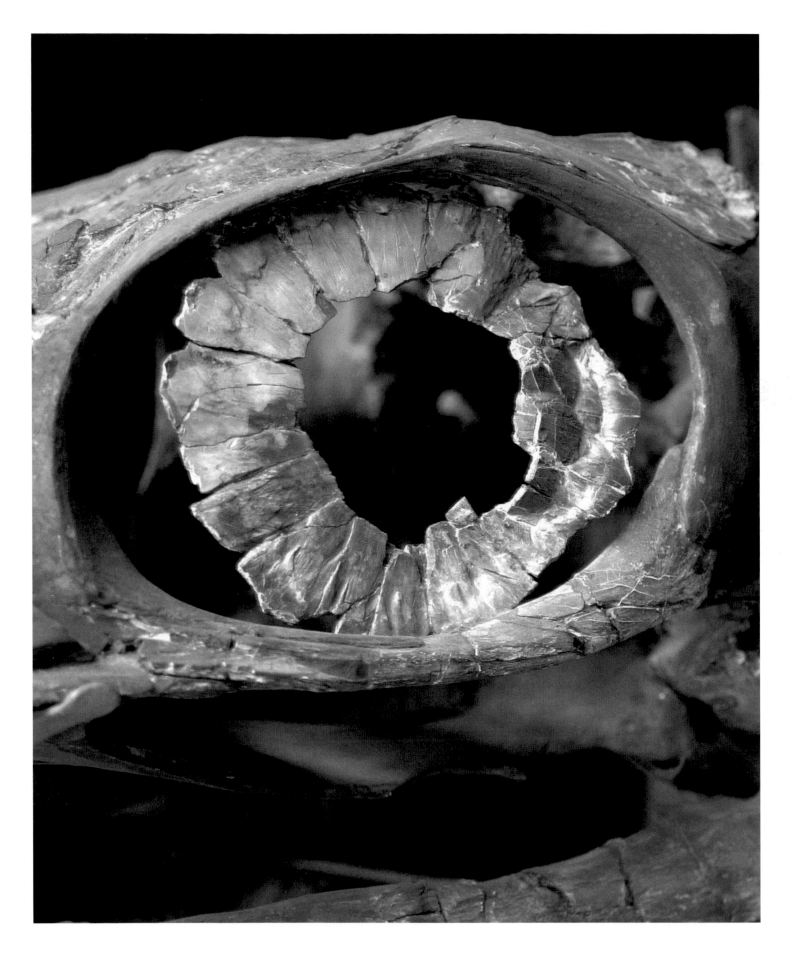

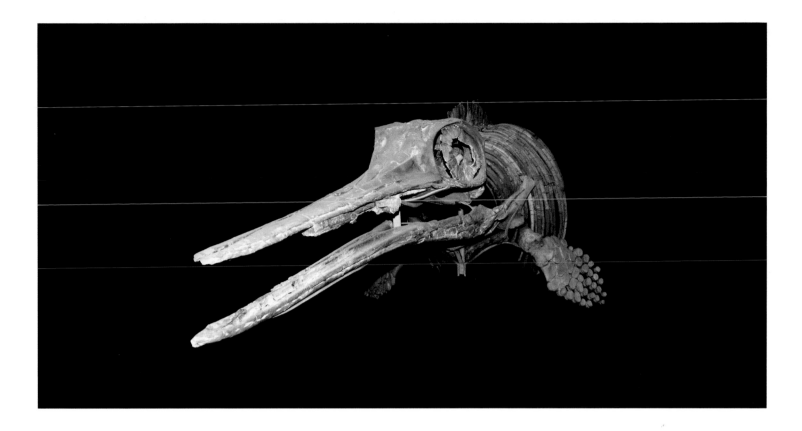

how big the aperture is in the middle, and so on – and so having bones that outline the size and proportions of a marine reptile's eye can offer palaeontologists clues about *Ophthalmosaurus*'s feeding habits.

Ophthalmosaurus has big eyes, large in both absolute size and relative to the animal's skull. In an adult about 6 metres (19 feet) long, the eyes were about 23 centimetres (9 inches) wide – nearly as wide as the eyes of other ichthyosaurs that were more than twice that length. Given that vision is such an important factor in finding prey and moving though the seas, the ichthyosaur's eyes must have been proportionally huge for a reason. Some experts suspect that that reason was diving deep to snatch small morsels from the depths.

The eyes of *Ophthalmosaurus* seem to be adapted to low-light conditions. Their overall size and the opening for light to reach the sensitive cells of the retina are big and resemble those of nocturnal species. Perhaps this means that *Ophthalmosaurus* hunted at night, pursuing nocturnal creatures that rose to feed in the upper zone of the ocean after darkness. One of the fossil beds in England where *Ophthalmosaurus* fossils have been found was probably about 50 metres (164 feet) deep, well within the Sunlight Zone. But where a fossil is buried doesn't necessarily reflect where that animal primarily lived or regularly visited, especially because

true deep-sea fossils are exceptionally rare. By looking beyond the eyes, palaeontologists have made a case that *Ophthalmosaurus* really did push the limits of an air-breathing reptile.

By comparing the eyes of *Ophthalmosaurus* to our understanding of how animals see, the ichthyosaur was probably able to distinguish moving prey at 300 metres (984 feet) or deeper. That's into the Twilight Zone, but deep enough that even the most brightly lit parts of the seas would start to darken. Furthermore, as the light fades, the more prevalent bioluminescence becomes – lights that might have guided these predators to prey. Likewise, drawing from the relationship between body mass and how long air-breathing animals can stay below the surface, *Ophthalmosaurus*, weighing around a ton (2,000 pounds) was probably able to dive for about 20 minutes – long enough for even a slowly cruising *Ophthalmosaurus* to reach 600 metres (1,968 feet) below the surface and come back up again. While other ichthyosaurs and marine reptiles likely stayed close to the surface, *Ophthalmosaurus* may have avoided competition by diving deep.

Opposite The eye contained fragile bones arranged in a ring. This might have helped prevent it from distorting under pressure when the reptile dived.

Above *Ophthalmosaurus* had a long snout and conical teeth.

Vampire Squid

IN THE ENTIRE HISTORY OF ZOOLOGY, THERE MAY NOT BE A MORE EVOCATIVE NAME THAN *VAMPYROTEUTHIS INFERNALIS*, THE "VAMPIRE SQUID FROM HELL". EVEN BETTER, THE TITLE IS AN IRONY. ONLY A THREAT TO SMALL PLANKTON, THE VAMPIRE SQUID LOOKS FAR MORE MENACING THAN IT TRULY IS.

The discovery of the vampire squid came during a time when oceanography was only just beginning to come into its own as a science. In the mid-nineteenth century, some ocean experts thought that there wasn't much life at all below 550 metres (1,804 feet). Trawls from greater depths turned up little, and so much of the ocean deep was thought to be azoic, or without life (see pages 50–51). The findings of the *Challenger* expedition in the 1870s (see pages 196–201) changed that picture, documenting many new deep-sea organisms and inspiring naturalists such as German researcher Carl Chun to go searching for even more clues about what lived far below the waves. In 1898–99, Chun was aboard the SS *Valdivia*, leading an expedition of the same name off Africa, when an odd cephalopod was brought to the surface.

Upon closer study, the cephalopod was extremely difficult to classify. It didn't seem to resemble any known species. In overall form, *Vampyroteuthis* looked like an octopus – more bulbous and less streamlined than the squid of the shallows. The cephalopod had eight arms with flexible, thorn-like projections down the

Right German naturalist Carl Chun was the first to scientifically describe a vampire squid.

Opposite Illustration of the vampire squid, based on Chun's work during the 1898–99 expedition.

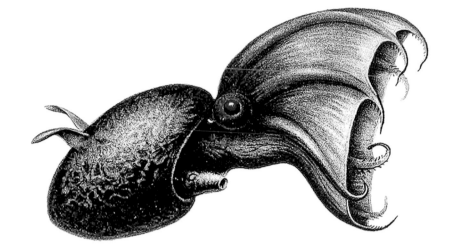

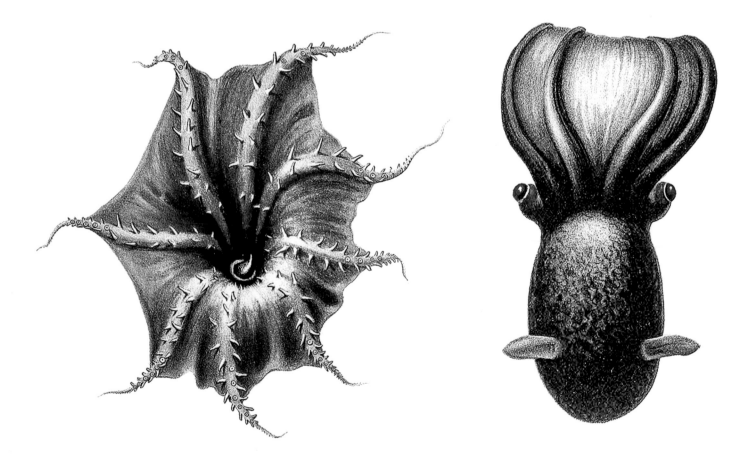

middle, yet lacked the specialized feeding tentacles that would be expected of a squid. But the vampire squid doesn't feed with its arms. Instead, in place of tentacles, *Vampyroteuthis* has two retractile threads covered in small filaments which help the animal detect and gather small prey and detritus that falls from above. Even though this unique creature is now recognized as being closer to octopus than to squid, it deviates so strongly from octopus anatomy that it is categorized within its own family – the Vampyroteuthidae.

As far as marine biologists know, there is only one living species of vampire squid. Once experts knew what to look for, however, they began to recognize vampire squid in the fossil record. While there is some debate, made all the more difficult by how rare well-preserved fossils of soft-bodied animals are, it seems that vampire squid have been around for at least 120 million years. One fossil,

more than 23 million years old, hints that vampire squid have been in deep, oxygen-depleted environments for at least that long. The cephalopods may have become adapted to low-oxygen environments documented in the deep past, and hung on in such harsh habitats ever since.

Today, vampire squid live in the ocean dark more than 600 metres (1,968 feet) below the surface. Even more specifically, and perhaps strangely, the cephalopods prefer a part of the ocean known as an oxygen minimum zone. These oxygen-depleted layers of the ocean usually occur between 200 and 1,000 metres (656–3,280 feet), and most organisms can't survive here. But vampire squid can. Within this zone, remote operated vehicles have been able to document how the vampire squid feed – floating along, their feeding filaments extended to hopefully capture whatever pieces of carrion or detritus they might touch. When the squid makes

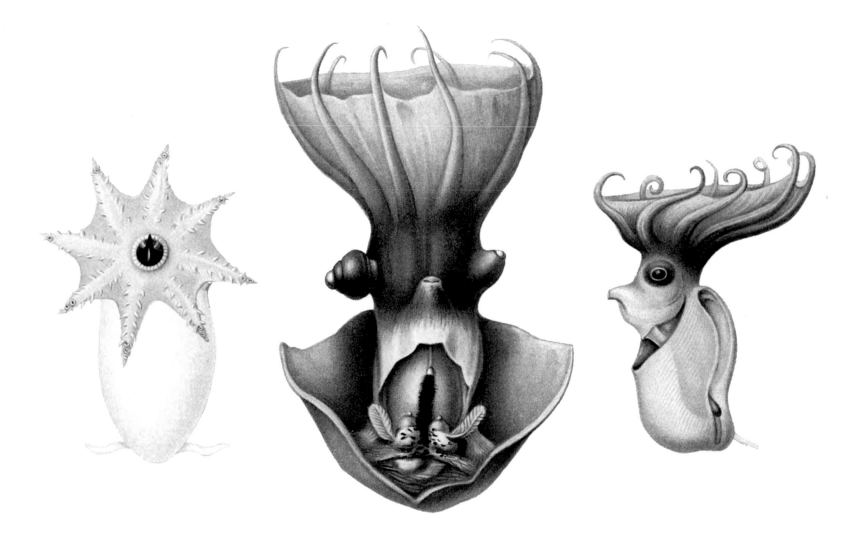

contact with food, it quickly moves towards it and then starts the process all over again. The vampire squid definitely lives life in the slow lane.

Of course, there are other creatures in the deep – and some of them are predators of the vampire squid. *Vampyroteuthis* does not move very fast, and even its swifter movements quickly drain the animal's stamina, so the cephalopod has two self-defence tactics. One is to use the bright photophores along the tips of its arms and at the base of each fin, setting off a disorienting light show that makes it harder to locate and catch the squid in the dark world of the deep sea. The other, which marine biologists call the pineapple posture, is to draw its arms up its body to present the predator with its exposed arms covered in spines. Those spines are soft and useless as actual defences, but they might help the squid look like too much trouble to eat.

How do vampire squid manage to procreate in such a distant habitat? What scientists expect mostly comes from other cephalopods, and might change as experts learn more. It's probably rare for vampire squid to encounter each other so when they do and choose to mate, a male passes a cylinder of sperm called a spermatophore to the female, which she stores in a special pouch until she's ready to fertilize her eggs and brood. When the squid hatch, they are as small as the plankton that the adults often feed on. To avoid becoming meals, the little ones go deeper to feed and grow until they can return to the waters they came from.

Opposite In life, the vampire squid is a deep red colour with a striking blue eye.

Above A great deal of what's known about vampire squid comes from dissections of specimens trawled to the surface.

Nautilus

MOST CEPHALOPODS LIVING IN TODAY'S SEAS ARE SOFT. OCTOPUS, SQUID, CUTTLEFISH, AND THEIR RELATIVES MAY HAVE INTERNAL PENS OR SUPPORT STRUCTURES, BUT ALMOST ALL LACK AN EXTERNAL SHELL. THAT'S NOTABLY DIFFERENT FROM MUCH OF THE DEEP PAST. FROM THE TIME THE FIRST CEPHALOPODS APPEARED AROUND 522 MILLION YEARS AGO, MANY FANTASTIC CEPHALOPOD LINEAGES EVOLVED ELABORATE AND HIGHLY ORNAMENTED SHELLS. TODAY, THOUGH, THERE IS ONLY ONE FORM OF CEPHALOPOD WITH HARD EXTERNAL SHELLS LIKE THEIR ANCIENT RELATIVES – NAUTILUS.

Even though we often talk about the nautilus, singular, there are actually six different species alive today. The most famous of all is the chambered nautilus, *Nautilus pompilius*, that bobs through the waters off Japan, Australia and Micronesia. Even by cephalopod standards, they look strange. The shell houses a series of chambers, each a little larger than the last, that act as the nautilus's home. Instead of sucker-lined arms, nautilus have up to 90 soft, flexible appendages called cirri that are housed in protective sheaths. And while the eyes of other cephalopods are some of the most complex and visually acute among animals, the eye of the nautilus is a simple pinhole arrangement that gives them a much blurrier view of the world. Despite jetting through our modern oceans, nautilus look just a touch prehistoric.

The very first nautilus species goes back to the Triassic, over 230 million years ago. The shelled invertebrates diversified and proliferated alongside their distant ammonite relatives, surviving millions of years as prey for the many marine saurians that evolved during the Age of Reptiles. But when a massive asteroid struck the planet 66 million years ago, ammonites went extinct and nautilus persisted. Precisely why such similar-looking cephalopods wouldn't both survive is a mystery. It might have something to do with differences in the way they reproduced, or changes to ocean acidity that ammonites couldn't cope with. Either way, by about 100,000 years after impact, nautilus were the only shell-covered cephalopods left.

Nautilus aren't born with such extravagant shells. They build them over time, throughout their whole lives. A newly hatched nautilus is only about 30 millimetres (1¼ inches) across, a part of the oceans' plankton. That's a difficult spot to be in, given that many marine species, from small fish to massive whales, feed on plankton. But the youngsters that survive will build their shells as they get bigger and mature. Nautilus live in the front-most chamber of their shells, bordered along the back by a septum. The shell keeps building itself with the life of the nautilus, and as each new chamber is added, the nautilus moves forward into the larger

Opposite Nautilus can survive to depths of about 703 metres (2,306 feet). Any lower and their shells would be crushed by the pressure.

space (the smaller chamber behind being added to the shell's coil). And these shells are sturdy. They maintain their shape to about 800 metres (2,624 feet) below the surface.

While many deep-sea creatures are world travellers, nautilus have a more restricted range. All living nautilus reside in the Indo-Pacific, often along the deep-water edges off continental slopes. They've been seen in water as deep as 703 metres (2,306 feet), though they are more commonly found at slightly shallower depths than that record. And as with the coelacanth, temperature is important to nautilus. Despite living in tropical waters, they stay deep enough to avoid temperatures above about 25°C (77°F).

A swimming nautilus might look a little ungainly. Gas within the shell helps nautilus alter their buoyancy, and they also move around by jet propulsion – expelling water to move themselves backwards through the water. Poor eyesight aside, nautilus cannot see where they're going when moving backwards.

Living nautilus have also taken another leaf from the deep-sea survival playbook. They are largely scavengers and opportunists. Rather than trying to snatch prey like an octopus or squid does,

nautilus most often feed on dead fish, the molted exoskeletons of crustaceans and other edible detritus scattered along the sea bottom. Even though nautilus in the ancient past were more varied and lived in a range of habitats and oceans, these cephalopods have survived by being part of the oceans' clean-up crew.

In many ways, nautilus seem more primitive than their other living relatives. That's not surprising given that their ancestors evolved about 100 million years before the last common ancestor of octopus and squid. Still, in lab experiments, researchers have found nautilus to have forms of both short-term and long-term memory that allow them to learn what a particular stimulus – such as a flash of light – means, and to tailor its response with experience. Instead of merely being a holdover from deep time, nautilus hold secrets to survival that we're only just beginning to perceive.

Opposite Living nautilus species are among the most archaic cephalopods, having eyes and arms that are very different from those of octopus, cuttlefish and squid.

Above X-ray of a nautilus shell shows the series of chambers.

Stromatolites

EVEN THOUGH LIFE ON EARTH IS CONSTANTLY EVOLVING, THERE ARE SOME
LIFE FORMS ON OUR PLANET THAT HAVE FOUND A COMFORTABLE NICHE
AND ARE STICKING WITH IT. THAT'S PART OF THE WONDER OF EVOLUTION
BY NATURAL SELECTION – IT EXPLAINS TRANSCENDENT CHANGE AS
WELL AS ORGANISMS THAT SEEM TO VARY LITTLE FROM THEIR ANCIENT
COUNTERPARTS. AMONG EARTH'S GREATEST STALWARTS ARE STROMATOLITES
– STRUCTURES THAT DATE BACK TO THE DAWN OF LIFE ON OUR PLANET.

Many stromatolites are found in shallow, sun-lit waters. Shark Bay in Australia, for example, is famous for the number of pillowy stromatolites visible along the shore. The proximity of stromatolites to the surface makes sense. These rocky mounds are not technically alive, but are instead precipitated out beneath colonies of photosynthetic cyanobacteria. At the start of the process, the cyanobacteria are spread over the surface of the sediment. As they photosynthesize and release oxygen, the cyanobacteria secrete sticky biochemicals that essentially glue the sediment together. But the cyanobacteria need to stay on top in order to survive. As they adhere bottom sediment particles together, the cyanobacteria begin to create little pedestals beneath themselves. Most recognizable stromatolites are these advanced colonies, their history glued together beneath them.

Stromatolites survive in very salty water. That's because lots of other organisms can eat them and break them down. Scientists hypothesize that the proliferation of stromatolites was curtailed with the origin of amoebas and foraminiferans (essentially amoebas with shells) that were capable of glomming on to the colonies and digesting them. Only those stromatolites that lived in waters too saline for the amoebas to reach were able to persist, and so the surviving structures only grow in a portion of their prehistoric range. Then again, researchers sometimes find stromatolites where they are not expecting them – like in the deep sea.

In 2018, geoscientists reported that stromatolites had been found in 731 metres (2,398 feet) of water in the Arabian Sea. Experts had not been expecting to find stromatolites so deep, well below the layers that sunlight penetrates. Yet there they were, slowly accreting in the deep.

While stromatolites close to the surface rely on photosynthesis, those deep down rely on a different biochemical pathway. The stromatolites found in the Arabian Sea are chemosynthetic, or formed with the help of microorganisms that do not require sunlight. By feeding off the methane seeping from the seafloor, these bacteria can carry out the same kind of growth and accretion as cyanobacteria do near the surface. It's a pathway that

Opposite Many modern stromatolites are found near the surface in extremely salty environments.

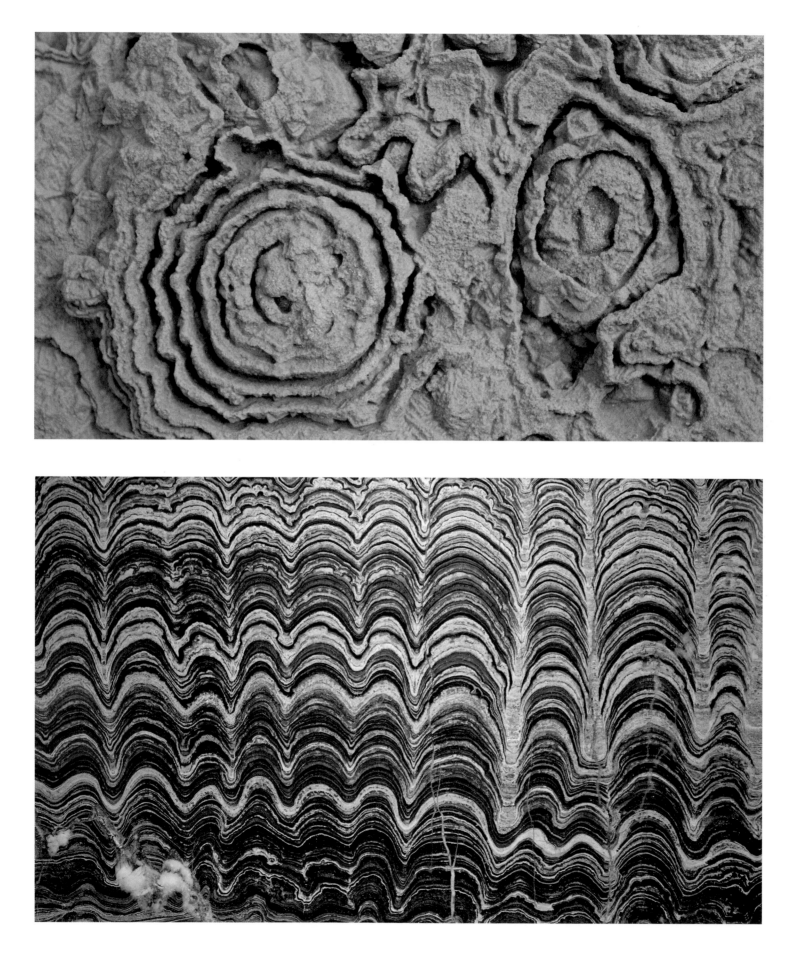

allows wildly different varieties of microorganisms to create very similar structures.

The discovery raises questions about the deep history of stromatolites. Fossil stromatolites, some of which date back billions of years, do not preserve the species of microorganisms that created them. Those mats of cyanobacteria – or whatever they were – are long gone, just leaving the form behind. What experts have been able to learn from modern stromatolites informs our understanding of the deep past. But while it's likely that many ancient stromatolites did indeed live in warm surface waters, there's also the distinct possibility that some fossil stromatolites were formed in other conditions – like that of the deep sea.

Chemosynthesis has only been known to researchers for less than a century, and observations of the phenomenon are even newer, dating to the 1970s. Ever since that time, though, scientists have wondered if this process might be related to the origin of life itself. If life did not require sunlight to originate, and instead could survive according to alternative pathways focused on breaking down compounds like methane that naturally flow out of the Earth, then it's possible that life didn't begin in a warm little pool on the shore, but in the darkness of the deep sea. And if that's the case, deep-sea stromatolites might offer a peek at what life

was like when it was a novelty on this planet. The fact that the waters the stromatolites were found in were oxygen-depleted, just like the waters of early Earth are thought to have been, seems to strengthen the connection. Likewise, researchers note, the internal structure of modern, photosynthesis-dependent stromatolites seem to differ from those of fossil stromatolites.

If chemosynthesis was more important to the formation of stromatolites than thought, the story of how our Earth came to be might need some major revisions. Perhaps photosynthetic bacteria were not as essential for oxygenating Earth's atmosphere as was previously thought, or maybe there was a greater number of pathways for stromatolites to form. Experts are only just beginning to dig into the possibilities, following the significance of these modern mounds back through eons upon eons.

Opposite above Ancient stromatolites were very similar to modern forms, down to the thinly layered bands.

Opposite below Cross-section through a stromatolite shows the bands.

Above Stromatolites on early Earth were probably more numerous before the evolution of organisms, like molluscs, that could eat the living colonies.

Following pages Stromatolites in Hamelin Pool in Western Australia are thought to be the oldest in the world.

Bathysphere

THE AGE OF OCEAN EXPLORATION IS RELATIVELY NEW. EVEN A
HUNDRED YEARS AGO, THERE WERE ONLY A LIMITED NUMBER OF
WAYS YOU COULD VISIT AND OBSERVE LIFE BENEATH THE WAVES. IN
THE DECADES BEFORE SCUBA CHANGED OUR RELATIONSHIP TO THE
OCEANS, THE ONLY WAY TO GET CONSIDERABLE BOTTOM TIME WAS
TO FIT INTO A BIG, CLUNKY, WATERPROOF SUIT FITTED WITH A HEAVY
HELMET CONNECTED TO THE SURFACE BY AN AIR HOSE. IT WAS
EITHER THAT OR SIMPLY HOLD YOUR BREATH IN A SHALLOW DIVE TO
SEE WHAT YOU COULD SEE.

But naturalist William Beebe was not content with what he could take in from the shallows. Oceanographers knew that there was life much further down, through organisms that either floated up to the surface in their last moments or were dredged from the bottom. There had to be a way to get down beyond the roughly 15-metre (50-foot) range divers were constrained by. So, with deeper waters in mind, researchers set about constructing a way to venture below the sunlight Zone – using an enclosed, underwater capsule called the Bathysphere.

Beebe had started making shallow dives with a homemade diving helmet in 1925. It didn't take long before he started wondering about how he might explore even deeper. By 1926, he was publicly speculating about visiting deeper waters in some kind of submersible vehicle – a disclosure that flooded his New York Zoological Park office with all sorts of design plans and sketches. One such proposal came from Otis Barton, a wealthy amateur naturalist, who came up with a design for a submersible and was willing to fund it. Beebe was initially skeptical, especially because he hated elaborate technology, but he was eventually won over by Barton's simple, spherical design. The Bathysphere would be a sealed ball that was lowered on a cable, supplied with air from above, as its occupants were lowered into the deep.

Perhaps such an expedition might seem quaint to us today, but Beebe, Barton, and their colleagues were taking a huge risk. There

Opposite Otis Barton inside the Bathysphere, which was only capable of carrying two people at a time to the deep.

was no escape hatch or even a safe way for the Bathysphere's occupants to return to the surface in case of a problem. If the pressure became too great for the sphere's design, then it might collapse and crush whoever was inside. The Bathysphere did not have a heating system either, and it was always possible that a malfunction or even an unwary animal might sever the air hose. Still, the two occupants who could fit inside would still be able to talk to the surface – relaying their condition as well as their observations – by way of a special telephone line running between the sphere and the support ship above.

All the research and preparations dictated that the iron Bathysphere had to be cast twice, as the first was too heavy for any ship to carry. Then Beebe and Barton ran several test dives, with crew and without, to make refinements and begin gathering information on the creatures they could view through the Bathysphere's portholes. A dive on 22 September 1932 was broadcast live by NBC Radio Network. Then, on 15 August 1934, Beebe and Barton made a record-breaking and historic dive over 922 metres (3,024 feet) down off Nonsuch Island, Bermuda. No one had witnessed the ocean at such depth, and even then the naturalists could only stay for about 5 minutes. The weight of the Bathysphere and its tether to the surface had to be treated with great respect; if the cable snapped, there would be no way to rescue or recover the crew inside. In fact, such a problem did occur. While the men were still deep below the surface, the rope that guided the steel cable on to its reel had severed and there was almost none left by the time they were back on board. Still, that didn't dissuade them and the Bathysphere made repeated dives.

Beebe took his explorations and observations to the popular press rather than scientific journals, a move that inspired contempt among colleagues. Then again, the Bathysphere was sealed and no specimens could be collected during the dives – only observed. Nevertheless, the Bathysphere and its explorations were famous enough to inspire the next generation of marine scientists – researchers who can stay down deeper, longer, and gather more information than Beebe could have ever dreamed.

Opposite Incapable of moving on its own, the Bathysphere required
a support ship to lower, raise and otherwise assist it.

Above Despite the cramped conditions, the Bathysphere allowed naturalists to
get some of the first direct observations of life below the Photic Zone.

Diel Vertical Migration

OVER DECADES OF EXPLORATION, OCEANOGRAPHERS HAVE IDENTIFIED PARTICULAR ZONES OF THE SEA AND REFINED DEPTHS WHERE SEAGOING CREATURES LIVE. ON PAPER, EVERYTHING CAN SEEM ORDERLY. BUT THE FACT OF THE MATTER IS THAT EARTH'S OCEANS ARE EXPANSIVE, DEEP, AND NOT SHARPLY DEFINED. MANY ORGANISMS DON'T REMAIN AT A SPECIFIC DEPTH, FINELY ATTUNED TO ITS PARTICULARS, BUT INSTEAD TRAVEL UP AND DOWN THE WATER COLUMN. IN FACT, SUCH MOVEMENT IS AN IMPORTANT PART OF OCEAN BIOLOGY.

Every day and every night, plankton and the creatures that feed on these tiny organisms rise and fall following the dark. Marine scientists know this as diel vertical migration, a regular pattern of the seas.

Naturalists weren't aware that such a pattern existed until the early nineteenth century. In 1817, the famed French anatomist Georges Cuvier noticed that tiny crustaceans called *Daphnia* could be found at the surface at some times and not others. The pattern wasn't because the *Daphnia* were sick or dying – like some giant squid or oarfish that come closer to the surface at the end of their lives – but was a real part of their daily cycle.

Experts didn't know just how many organisms made this daily migration. It wasn't just *Daphnia*. During the Second World War, for example, US Navy experiments with sonar kept detecting strange reverberations in the deep sea. While military minds fretted that these signals were being created by enemy submarines, researchers from the Scripps Institution of Oceanography discovered that the sonar was picking up dense pockets of plankton that lived at depth. And once marine scientists became aware of this layer, they found that it moved. Many, many species

of plankton migrated up and down the water column according to light levels – as did some creatures that fed on the plankton.

Naturally, the oceans are full of organisms belonging to so many different kingdoms, phyla, classes, and more that there is no single great migration. There are several forms of migration made up of varied species.

The most common and most well-known form of the phenomenon is the daily nocturnal vertical migration. Deep-sea plankton and their predators stay in the deep and dark during the day, but rise towards the surface during dusk before sinking again with the dawn. This affects large creatures as well as small ones – the megamouth shark (see pages 40–43) feeds on plankton

Opposite The daily migration of marine organisms was accidentally discovered by the US military using sonar.

Below *Daphia* provided the first evidence that some sea creatures migrate daily through the water column.

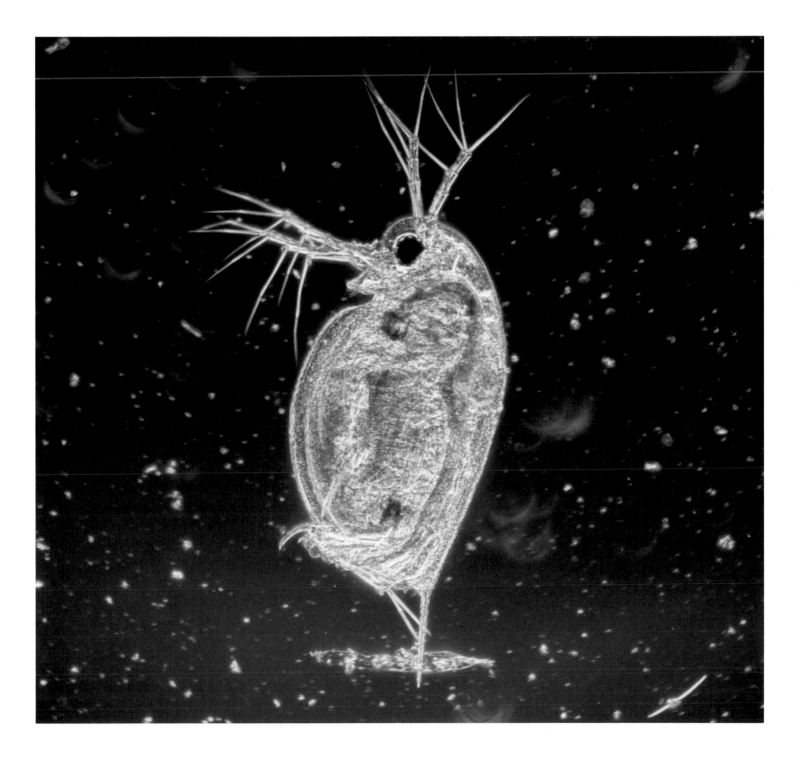

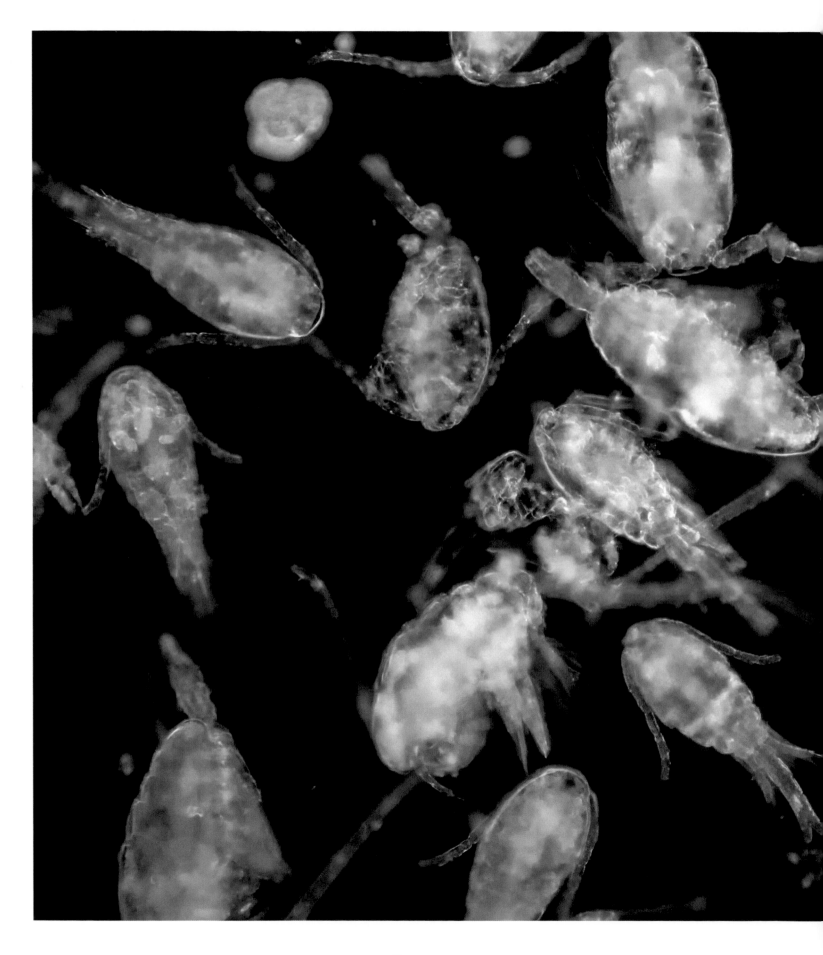

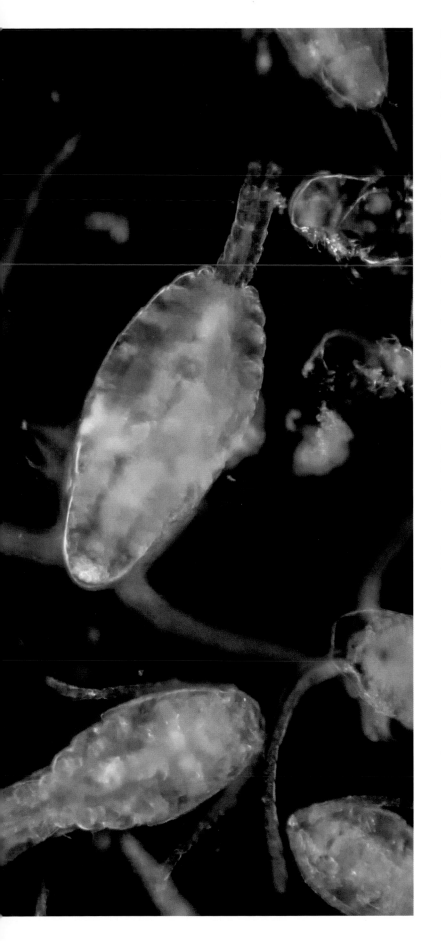

and so follows their migration pattern. Other organisms rise during the day rather than at night, some might rise and sink according to the seasons, and some rise in the evening, sink, rise again in the morning, and sink again to make two trips in a day. Some organisms – such as giant spider crab larvae (see pages 58–61) or hatchlings of giant oarfish (see pages 100–103) – spend the earliest parts of their lives closer to the surface before going lower as they increase in size. For one reason or another, the oceans' plankton are almost always in motion through the layers of the seas.

But how do these organisms know when to rise through the water column and when to sink? It may not be a sense of knowing so much as responding to particular cues. Some cues come from the environment and are part of the oceans' daily rhythms, like changes in light or shifts in temperature. Organisms may simply follow the conditions where they feel most comfortable. But there are likely some internal triggers for these daily migrations, too. Experiments with tiny crustaceans called copepods have found that these tiny animals continue to migrate up and down through the water column each day when they are kept in total darkness in the lab, and a few other organisms have been shown to do the same – moving regardless of the light. It might, in part, be their biological clock that makes them rise and sink no matter the surrounding conditions. In other cases, small animals may seek refuge at different depths – where predation is lessened – and will change where they live as they become larger.

The various forms of migration are critical to ocean health. Life in the deep ocean is largely fed by detritus and decaying organic matter that falls from the upper zones. If all those nutrients just stayed at the depths, then life in the higher zones might be sparser. But in addition to animals – like some whales – that feed at great depth and excrete some of the nutrients back near the surface (see pages 20–25), deep-dwelling plankton that migrate upwards also become food for organisms that dwell in those upper layers. This interaction helps return some of the nutrients that fall to the deep sea back to creatures in the Sunlight Zone. Even though the nature of these migrations is still mysterious, we know that they are essential to the wellbeing of the ocean.

Left Copepods and other plankton form the foundation of ocean ecosystems, so many animals that feed on plankton follow them up and down between night and day.

Goblin Shark

NOT ALL STRANGE FISH ARE FIRST SEEN IN THE DEPTHS. THE WORLD'S FOSSIL RECORD RETAINS PATCHY EVIDENCE OF THE DEEP THROUGH THE AGES, AND SOMETIMES SCIENTISTS FIND THE FOSSILS OF STRANGE CREATURES BEFORE THEY REALIZE SUCH SPECIES ARE STILL WITH US. THE SLEEK, STRANGE GOBLIN SHARK THAT SWISHES THROUGH THE MODERN-DAY DEEP IS ONE SUCH CASE.

In 1887, British naturalist and fossil fish expert James William Davis described a strange fossil shark found in Sahel Alma, Lebanon. It was a stunning specimen, complete from nose to tail and with an outline of the shark's body – much more informative than the handfuls of fossil teeth that many ancient sharks are solely represented by. Davis named the shark *Scapanorhynchus lewisii*, a fish that had lived millions of years ago and was presumed extinct.

Just over a decade later, in 1898, the American ichthyologist David Starr Jordan studied an odd shark that had been caught in Japan's Sagami Bay. The fish didn't look quite like anything anyone had seen before. The shark had a long, tapering body and a flat, shovel-like snout that left its needle-toothed jaws hanging below. Jordan named the shark *Mitsukurina owstoni*; the common name is a translation of its Japanese name, *tenguzame*, reflecting the shark's likeness to Japan's mythological, long-nosed tengu.

Other naturalists soon recognized the resemblance between the fossil shark and the goblin shark caught in Sagami Bay. In fact, some experts even thought that they might be the same genus – an example, such as the duck-billed platypus or the coelacanth, of a creature seemingly changing little over time. The specifics of that idea were eventually discounted, but it's

still true that goblin sharks – like today's *Mitsukurina* – have been around for a very long time, since before the days of *Tyrannosaurus* and *Triceratops*.

Like many other large, carnivorous members of the deep-sea community, goblin sharks live in deep water off the world's continental shelves. Goblin sharks have been observed or caught off the coasts of North and South America, Africa, Europe, Asia and Australia, often at depths between 200 and 1,000 metres (656–3,280 feet), although some go deeper. A goblin shark tooth was once found embedded in an undersea cable at about 1,370 metres (4,494 feet) down.

Why a goblin shark might try to bite a chunk out of an undersea cable might have something to do with how sensitive

Opposite above While many photos of goblin sharks show them with their jaws extended, most of the time the shark's jaws fit neatly beneath its long snout.

Opposite below left Naturalist David Starr Jordan described the goblin shark about a decade after the first fossil shark was found.

Opposite below right The tengu – a creature from Japanese folklore – provided inspiration for the shark's common name.

this deep-sea shark is. In addition to senses like taste and touch, sharks have a special electromagnetic sense all their own. Jelly-filled pores called ampullae of Lorenzini run along the snout of the goblin shark, as well as sharks in general, and help the fish detect the weak electrical fields that living things emit. The electricity running through an undersea cable might have intrigued or irritated the shark – a problem that companies such as Google still have to contend with, as deep-sea sharks have bitten telecommunications cables so fiercely that it's led to internet outages.

Like many other sharks, the goblin shark bites by swinging its jaws forward and away from the base of its skull to better grip and snatch its preferred meals of rattails and dragonfishes. But this shark has taken the feeding method to an extreme, in what may be an adaptation to life as an ambush predator that can certainly make the most of some extra reach.

While the shark's jaws are snugly slotted into its long snout most of the time, making its head look like a long paddle, those jaws can jut forward in an instant. The shark has special ligaments that are usually kept under tension as the shark swims around in the dark. When the goblin shark detects prey, those ligaments relax and, as with a rubber band, they help the lower jaw spring forward extremely fast, with the upper jaws close behind. Many dead and preserved specimens of goblin sharks demonstrate this arrangement, which sometimes makes the shark look more grotesque than it truly is.

No one really knows how many goblin sharks there are in the deep sea. The shark is so rarely caught, in an environment so far away from humans, that it is listed as being of little concern to conservation groups. But these fish may be highly sensitive. In 2003, over 100 goblin sharks were caught off Taiwan. No one is sure why. Some experts think an earthquake just before the record catches might have disturbed the fish. Goblin sharks have been swimming the seas for 100 million years, but we still know remarkably little about them.

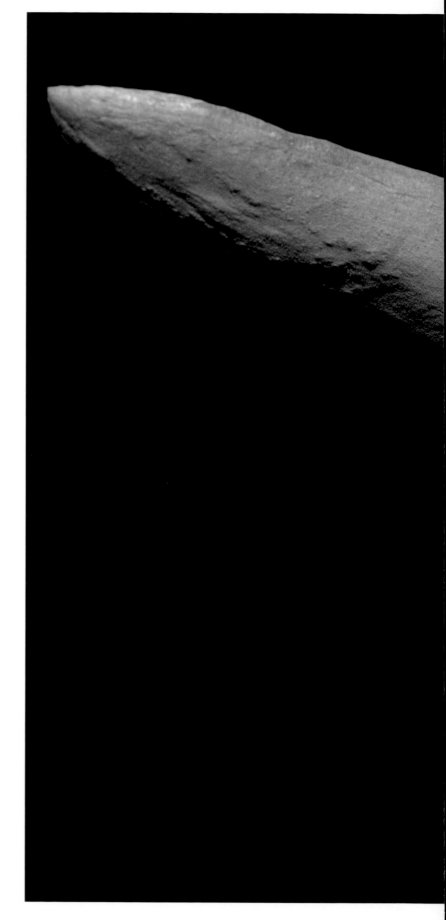

Right Goblin sharks have thin, needle-like teeth best suited to nabbing slippery prey like small fish and squid.

Giant Squid

THE CREATURE WASN'T SUPPOSED TO EXIST. THE CREATURE THAT NINETEENTH-CENTURY NATURALIST JAPETUS STEENSTRUP SET ABOUT DESCRIBING SEEMED LIKE THE CLOSEST THING SCIENCE HAD FOUND TO A LIVING LEGEND: A GIANT SQUID.

By the time of Steenstrup's study in 1857, the Age of Exploration was proving sensational tales of sea serpents and gigantic monsters to be unfounded. All the creatures that were supposed to dwell on the margins of nautical maps were either being brought into the realm of scientific inquiry or dismissed as myth. And yet, time and again, stories persisted of enormous squid washing up on beaches. Ship captains sighted odd creatures near the surface that seemed awfully squid-like. By chance, Steenstrup had been able to acquire the beak of a gigantic cephalopod that had washed ashore. That beak was the key. Such a specimen, made of two keratinous halves that came together to nip and punch at prey, was the tangible proof that gargantuan cephalopods truly dwell in the deep. "From all evidences," he wrote, "the stranded animal must thus belong not only to the large, but to the really gigantic cephalopods, whose existence has on the whole been doubted."

Steenstrup gave this mythical creature a name, *Architeuthis dux*. And even if other experts doubted Steenstrup's conclusions, further proof of the enormous animal soon appeared. In 1861 the French war ship *Alecton* was plying the waters near the Canary Islands when the crew happened upon a moribund giant squid floating at the surface. This was a truly enormous *poulpe* (octopus), so big that attempts to haul the entire animal aboard caused the squid's many-armed head to be separated from the rest of its body. All the same, there could no longer be any doubt. Somewhere out there, in the forbidding sea, there dwelled squid far larger than any found in a fish market.

Despite sparking the imaginations of scientists, sailors and the general public, the giant squid largely remained a mystery for decades after Steenstrup gave the squishy animal a scientific name. Almost everything naturalists knew about the animal came from corpses that washed ashore, were found dying near the surface, or were reduced to sucker hooks and beaks in the stomachs of slaughtered sperm whales. No one knew what the giant squid's natural habitat was because no one had seen a healthy one alive, much less anything else that couldn't be gleaned from decomposing carcasses. The squid seemed to live in every ocean, and bodies kept floating up to the surface, but even as marine biologists began to use remote operated vehicles, submersibles, and even cameras attached to squid-hungry whales, seemingly no one could catch a glimpse of *Architeuthis*.

It took until 2005 to get the first pictures of a living giant squid in its oceanic home. In that year, marine biologists Tsunemi Kubodera and Kyoichi Mori debuted the first still images of a giant squid they had photographed trying to remove some bait from a line the researchers had slipped down to a depth of 900 metres (2,952 feet) below the surface. The fact that the squid actively tried to snag a snack – accidentally leaving a tentacle behind in the process – was useful information by itself. Biologists weren't

Opposite The crew of the *Alecton* attempt to haul the body of a giant squid on board.

sure how the squid caught their meals. Some thought they were active predators, like many of their smaller relatives, while others proposed that giant squid might drift along in deep-sea currents with their tentacles extended, waiting for an unlucky morsel to blunder along. The snapshots showed that squid actively went in search of their meals, scavenging when necessary.

Researchers didn't have to wait long for additional information about the giant squid to surface. Other researchers from spots around the world began to successfully locate and film living giant squid. In 2013, marine biologists at long last recorded video of a live giant squid in its natural habitat off Japan's Ogasawara archipelago in about 700 metres (2,296 feet) of water. In 2019, too, footage taken in the Gulf of Mexico recorded a visit from a giant squid – which briefly wrapped its arms around the Medusa camera system used by the researchers.

There's still a great deal that remains mysterious about giant squid, but biologists have come to know a few things.

These cephalopods can reach lengths of 12 metres (39 feet) or more, and they live in every ocean at depths between 300 and 1,000 metres (984–3,280 feet). While some researchers used to speculate that there were multiple giant squid species, genetic analysis indicates that there is only one, global giant squid – *Architeuthis dux*. Gut contents indicate that giant squid eat fish and other, smaller squid, which they spot thanks to the largest eyes in the animal kingdom. Nevertheless, no matter how large the squid looms in our imaginations, we barely know this celebrity cephalopod.

Above Most of what we know about giant squid comes from carcasses that wash ashore.

Opposite above Chart of northern Europe, 1539, with sea monsters.

Opposite below Dr Tsunemi Kubodera shows the first still images of a giant squid on his laptop, 2005.

Following pages An injured giant squid floats at the surface.

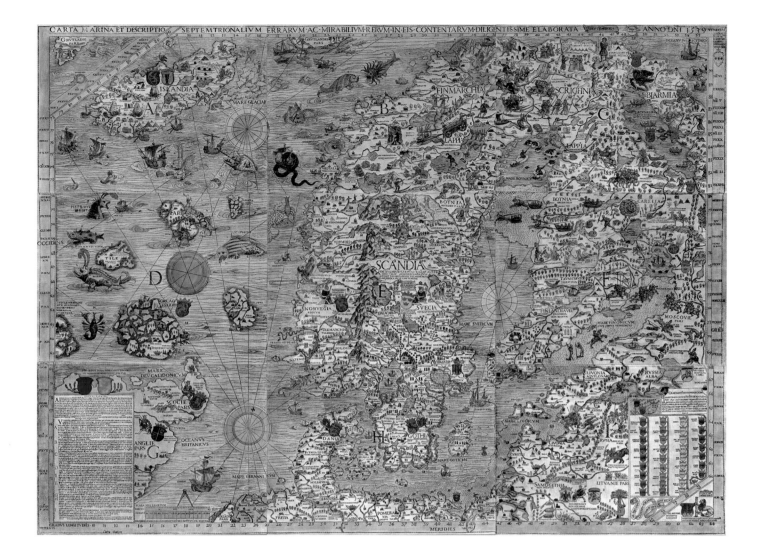

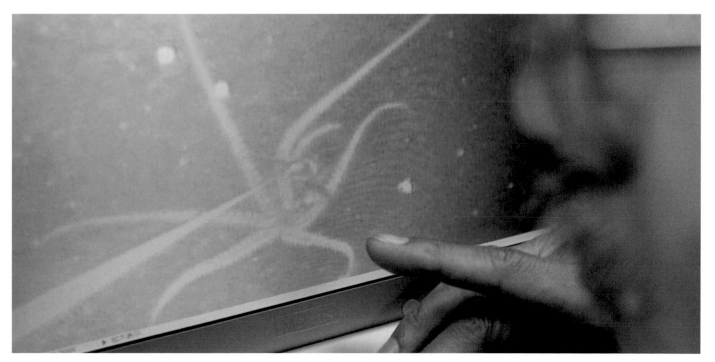

Cookie-Cutter Shark

SHARKS ARE WELL KNOWN FOR THEIR BIG BITES. LARGE PARTS OF THEIR ANATOMY AND BEHAVIOUR ARE BASED AROUND THE WAYS IN WHICH THESE CARTILAGINOUS FISH CAPTURE AND CONSUME PREY. AND GIVEN THE HUNDREDS OF SHARK SPECIES KNOWN, IT'S NO SURPRISE THAT SOME OF THESE SPECIES TAKE THEIR SPECIALIZATIONS TO EXTREMES. WHILE ONLY ABOUT 50 CENTIMETRES (19 INCHES) IN LENGTH, THE COOKIE-CUTTER SHARK HAS A BITE THAT MAKES A GREAT WHITE LOOK LIKE AN AMATEUR.

The scientific discovery of the cookie-cutter shark – also known as the cigar shark – goes all the way back to 1824. French anatomists Jean René Constant Quoy and Joseph Paul Gaimard had described the small and unusual shark as part of a 13-volume report on the scientific findings of the 1817–20 voyage of the *Uranie*. They originally called the shark *"Scymnus" brasiliensis* for the place it was caught, off the coast of Brazil, with an American ichthyologist revising the name to *Isistius brasiliensis* some years later.

But the true nature of the shark wouldn't become clear for over a century and a half. The shark spends the day in the deep, over 1,000 metres (3,280 feet) down, and comes much closer to the surface at night. Given that the fish spent almost all of its time in the darkness and was practically impossible to observe in life, no one really knew how it sustained itself – or its connection to strange wounds found on various sea creatures.

Despite a great deal of time spent at depth, the cookie-cutter shark seems to prefer warmer latitudes in the world's seas around the equator. And in these waters, larger fish and dolphins could be seen with odd punctures in their flesh. In fact, these wounds were the inspiration for a Samoan story that skipjack tuna would leave portions of their flesh as sacrifices when they came into Palauli Bay. In time, marine biologists picked up on the phenomenon, too. Was it some kind of bacterial infection, or perhaps a parasite? No one could be quite sure what was creating the crescent-shape wounds that looked like some sort of bite.

The picture came together slowly. In 1963, ichthyologist Donald Strasburg reported that *Isistius brasiliensis* does not shed teeth one at a time like most sharks do, but instead replaced its entire lower tooth row all at once as if it were some kind of razor cartridge. Strasburg wondered what kind of lifestyle could necessitate such a strange adaptation that allowed the little shark to always have a perfectly even set of large, sharp teeth. The answer would come in 1969, during a strange moment of discovery aboard the R/V *Townsend Cromwell* as it trawled the Pacific for deep-sea fish.

The nighttime trawls brought dead and dying cigar sharks to the surface. Marine biologist Everet Jones was on board and recalled Strasburg's question about the shark's teeth and the

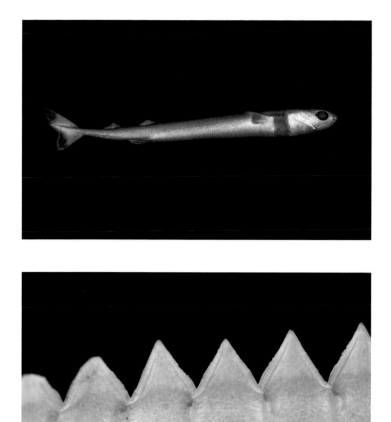

possible connection to the mysterious wounds on fish and whales. Jones remarked on the idea to research assistant John Fowler, who then took a nearly dead cigar shark and pressed its mouth against the side of a dead fish to see if the shark might bite with a reflex action. The shark did, and scooped out a crescent-shaped wound just like the ones that had so mystified scientists. The researchers had their culprit, and closer inspection of photographed wounds on tuna, whales and swordfish matched up with the teeth of the cookie-cutter.

Marine biologists have learned a great deal more about the cookie-cutter shark since that first nighttime experiment. These sharks are not especially strong swimmers, and, alone or in small schools, practically hover in the water column as they wait for large prey to pass by them. When the cookiecutter strikes, the shark uses a specialized set of lips to briefly suck on to the surface of its prey and quickly bite down to scoop out a large chunk of skin, blubber, muscle, or whatever else it can get a mouthful of. Zoologists call this way of feeding facultative ectoparasitism, to describe when an organism is specialized to rely on part of

another for its food without making a permanent attachment to or killing the host. This makes the cookie-cutter shark a kind of flesh grazer, taking chunks out of various animals but not causing debilitating injury in the process.

Naturally, the way the shark feeds has been its biggest claim to fame. But that's hardly all. The cookie-cutter shark is also bioluminescent, and, in fact, might have the most intense underwater glow of any known shark. The photophores that create light on the shark's body – a kind of camouflage that disrupts the shark's silhouette against would-be predators – can continue to shine for hours after death. In habitats where light rarely reaches, a little glow can help disguise the cookie-cutter.

Top left The cookie-cutter shark is also known as the "cigar shark" for its small size.

Above left The triangular teeth fit in an almost razor-like arrangement and are shed all at once instead of one by one.

Above right The formidable bite of the little cookie-cutter shark.

1,000 metres

Giant Oarfish

IN MID-OCTOBER 2013, A STRANGE FISH WASHED UP ON OCEANSIDE HARBOR BEACH ON THE CALIFORNIA COAST. ALMOST 4 METRES (13 FEET) LONG, THE SINUOUS CREATURE WAS QUICKLY IDENTIFIED AS A GIANT OARFISH – THE LONGEST BONY FISH IN THE SEA. THOUGHT TO BE THE INSPIRATION FOR SOME TALL TALES ABOUT SEA SERPENTS, THE FISH IS SO BIG AND SO RARELY SEEN THAT IT'S A LIVING LEGEND.

There are at least two species of oarfish in today's oceans. The smaller of the two, the streamer fish, lives at depths below below 500 metres (1,640 feet) and has a ribbon-like ornament jutting from its head. But it's the giant oarfish that often makes headlines when it's seen, a huge and sinuous fish that is still poorly known.

Called *Regalecus glesne* by specialists, the giant oarfish is an open-water species. The largest verified individuals get to be 8 metres (26 feet) long, though there are rumours that 11-metre (36-foot) oarfish have been seen, as well. The individuals that wash up on shore are actually far from their usual home. These fish live deep below the surface of the open sea, up to 1,000 metres (3,280 feet) down and across a broad range of the world's tropic and semitropic oceans. The fish's preferred depth – like many deep-sea organisms – makes it challenging to study. Despite being named in 1772, the first camera footage of a giant oarfish in its home habitat didn't surface until 2010.

The fish is so seldom seen that marine biologists aren't even entirely sure how the sinuous giant moves through the water column. One oarfish spotted in the Bahamas kept its body straight and propelled itself by undulating the long, red dorsal fin atop its back. This is called amiiform locomotion, similar to that used by freshwater bowfin fish. Then again, giant oarfish have also been seen swimming vertically, perpendicular to the ocean surface. Precisely why they move this way is unclear, but when you spend your entire life underwater with thousands of metres/feet of depth, there are many different directions and ways to move around.

Much of what marine biologists have been able to discern about giant oarfish comes from individuals that become stranded or wash ashore. The fish is toothless, with more than 40 gill rakers inside its throat to help filter out small plankton and other morsels from the water column. Gut contents from stranded fish indicate that they eat shrimp, jellyfish, small squid and other tiny prey. And despite what you might think for a fish of such length, most vital organs in the giant oarfish are crammed close by the head and so the majority of the fish's body is tail. This might be a defensive adaptation. If a shark or other ocean predator were to attack, the oarfish would be able to survive if it lost a little bit of its tail.

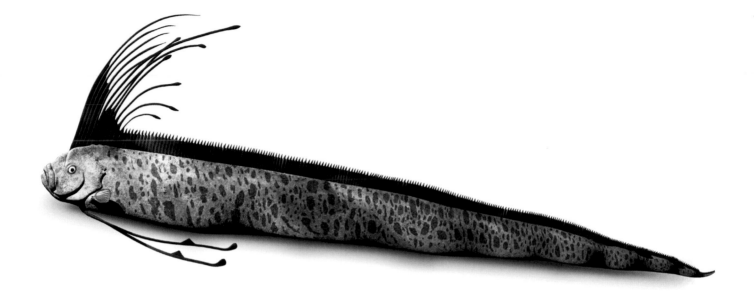

Top Giant oarfish are immediately recognizable from the bright red
"mane" running along their backs.

Above Giant oarfish live too deep to be found easily. Marine biologists
have had to learn a great deal from stranded specimens, like this fish
discovered by a US Navy installation in 1996.

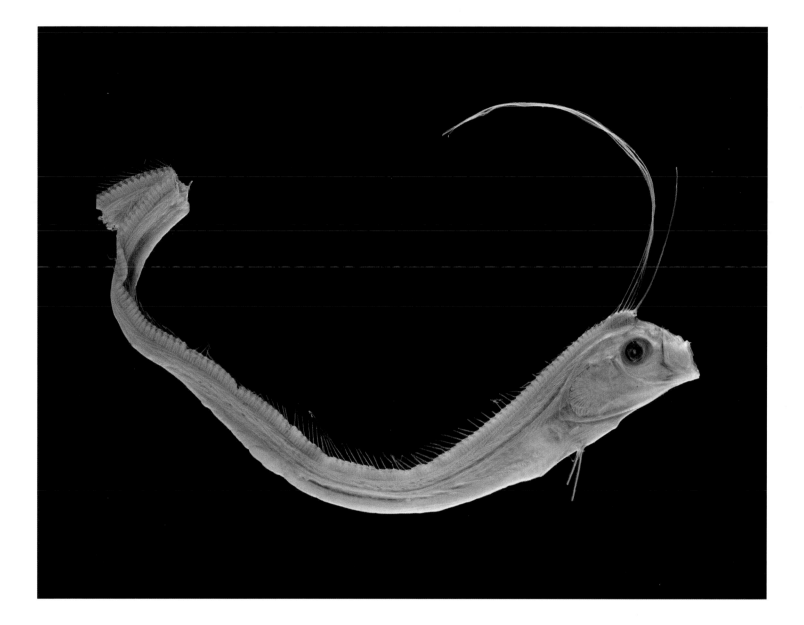

Despite the potential dangers, the giant oarfish is best adapted to life in the deep sea. Oarfish seen near the surface, despite making a striking impression upon any people who might spot them, are often sick or dying. The surface waters are a harsh place for these fish. Down deep, there are no strong currents to buffet the fish about and so they don't develop the kind of sturdy musculature that would allow them to navigate the more turbulent waters near the surface. The relatively calm and consistent conditions through the deep sea, across latitudes and longitudes, may be what allow this fish to range worldwide, but at restricted depths.

But these ocean giants start off life very small. When giant oarfish spawn, which may be several times over the course of a month or two, the fertilized eggs from the female fish drift out into the ocean plankton. From there, it's a matter of luck. Many of the eggs will be eaten by other creatures that inhabit the deep ocean. But the few that manage to hatch will be tiny ribbons that dwell just below the surface, where they can suck up detritus and plankton smaller than themselves. They swim with long pectoral fins – not the iconic dorsal fin seen in adults – and their mouths stay open to take in any nutritious specks they need to grow. It's only as the oarfish mature and become larger that they begin to go deeper, and are rarely seen at the surface again.

Opposite Gut contents from giant oarfish washed ashore often contain jellyfish, squid and shrimp.

Above Juvenile oarfish spend much of their time near the surface feeding on plankton, only inhabiting greater depths as they grow larger.

Lanternfish

SONAR TECHNICIANS WERE CONFUSED BY THE READINGS. DURING
THE SECOND WORLD WAR, AS NAVIES BEGAN TO USE SOUND
TO NAVIGATE THE DEEP SEA AND DETECT SUBMARINES, THE
OPERATORS LISTENING IN KEPT FINDING THAT THE SEAFLOOR
WAS IN THE WRONG PLACE – AND THAT IT MOVED. IN DAYLIGHT
HOURS THE SEA BOTTOM SEEMED TO BE UP TO 500 METRES (1,640
FEET) DEEP, YET AT NIGHT THE SAME READINGS WERE COMING UP
CLOSER TO THE SURFACE.

Both seemed wrong compared to oceanic surveys, until researchers realized that what they were reading as the seafloor wasn't really the bottom at all. They had found the deep scattering layer – not so much a level of the ocean but of innumerable small creatures, including millions and millions of fish. The air-filled swim bladders of these little fish were causing the sonar signals to be misread, and many of these tiny, confounding swimmers were lanternfish.

Known to experts as myctophids, lanternfish get their name from the fact that these fish bioluminesce. The placement of their glowing tissues varies from species to species, ranging from the underside of the belly to the tip of the snout. And there are a lot of these fish. Marine biologists have counted more than 246 species since the late nineteenth century, ranging from 2 to 30 centimetres (¾ to 11¾ inches), and these fish might be the most numerous vertebrates on the planet. Experts estimate that 65 per cent of deep-sea fish biomass is lanternfish – more than 550 million tonnes of these sleek little swimmers.

Lanternfish have been around for a very long time. Tiny, hard structures that aid fish in hearing and balance – called otoliths – have allowed palaeontologists to track the origin of these fish back to more than 55 million years ago. Back then, the ancestors of today's lanternfish didn't dwell in deep water. Their remains are found over the continental shelf, in relatively shallow water. It wasn't until about 33 million years ago that lanternfish started going deeper, their otoliths appearing in deeper sediments. Even then, these fish weren't behaving like their modern counterparts. The daily migration of lanternfish from deep to shallower waters didn't start until about 15 million years ago, a time when changes to nutrient cycling in

Opposite Lanternfish get their name from the prominent bioluminescent organs along their bodies.

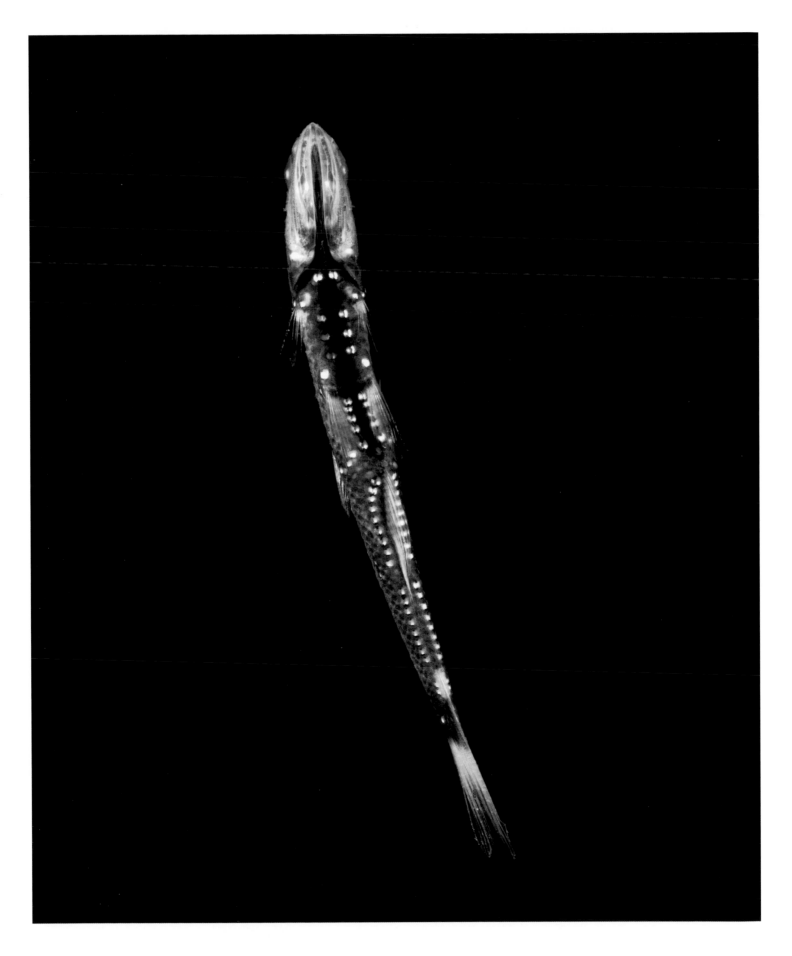

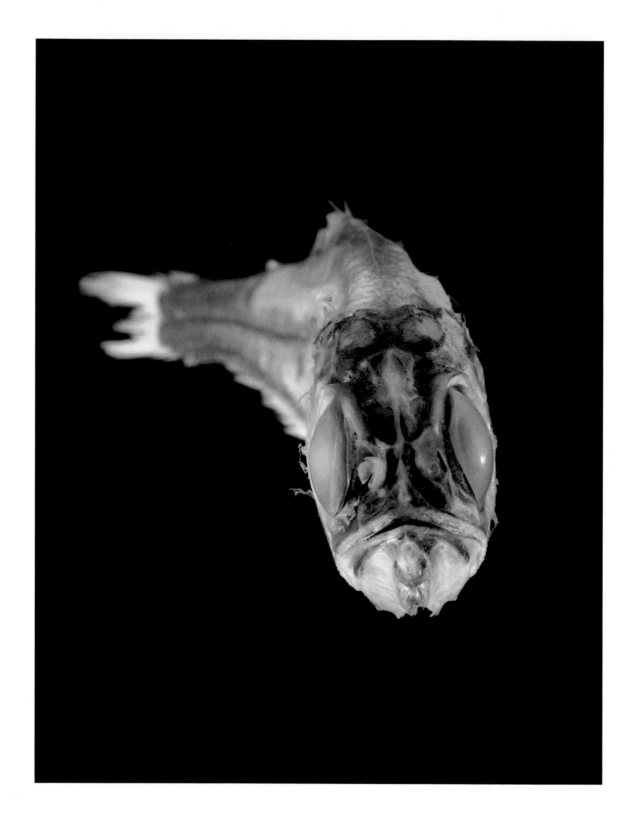

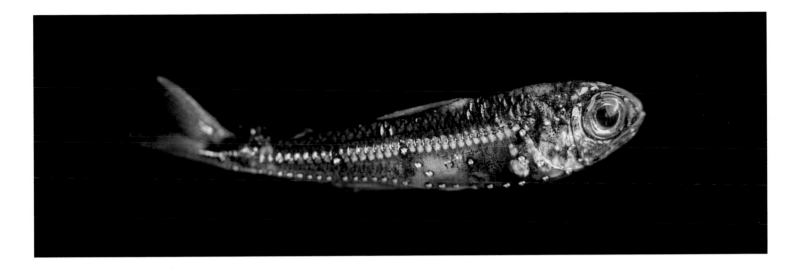

the oceans allowed plankton to bloom and provide ample food for creatures ranging from whales to the tiny lanternfish.

Much like the megamouth shark (see pages 40–43) and other deep-sea creatures dependent on plankton, lanternfish spend daylight hours down deep. Many lanternfish swim through the dark of the Twilight Zone and into the Midnight Zone, between 300 and 1,500 metres (984–4,921 feet) below. But because these fish feed on tiny zooplankton, they need to follow the food. As their preferred morsels rise towards the surface at dusk, the lanternfish follow. This migration also means the small and abundant fish have less risk of becoming meals themselves. Not that all lanternfish follow this pattern. Some reside in the deep for almost their entire lives. And it iss these differences that may have spurred lanternfish to evolve their characteristic illumination.

If lanternfish only used their bioluminescence to blend into the sea or avoid predators, then we might expect most species to have similar arrangements of photophores to best distract or deceive. But that's not the case. *Diaphus* lanternfish have photophores near their eyes, often likened to headlights, while others have glowing spots near their fin bases or under their bellies. The light can be yellow, blue or green, and sometimes the details of their light patterns differ between males and females of the same species. All this diversity is an important clue. While some photophore

placements are likely to act as camouflage – causing the fish to look darker from above and lighter from below – others help these fish keep to schools of their own species or allow the fish to find mates in the darkness.

In fact, having such a unique and malleable way to communicate might explain why there are so many species of lanternfish compared to other deep-sea fish species. The deep sea is more like a void than any environment on land. There aren't as many geographical barriers that cause populations to become divided and evolve in different ways. But the way lanternfish communicate with each other allows for other evolutionary pressures – like sexual selection – to play a role and generate a greater number of unique species. When lanternfish want to attract or communicate with members of their own species, their light can be seen from about 30 metres (98 feet) away, a form of deep-sea advertisement that these swimmers have mastered. If you see a light in the ocean deep, chances are it's a lanternfish.

Opposite Lanternfish are considered to be one of the most common deep-sea creatures, making up about 65 per cent of all the deep-sea fish biomass.

Above Different lanternfish species have various arrangements of photophores, suggesting that some of these bioluminescent organs are for communication as well as camouflage.

Big Red Jelly

SOMETIMES NEW DEEP-SEA SPECIES ARE HIDING IN PLAIN
SIGHT. THAT'S HOW THE BIG RED JELLY *TIBURONIA GRANROJO*
WAS DISCOVERED – NOT THROUGH IMMEDIATE RECOGNITION
BUT THROUGH CAREFUL SCIENTIFIC DETECTIVE WORK.

In 1998, marine biologist George Matsumoto got a call about a strange jellyfish that had been seen on an underwater geology expedition. This isn't all that unusual. Scientists with varied backgrounds study and explore the deep sea, and sometimes a crew looking at one phenomenon – such as the geology of the ocean floor – spot unusual creatures or phenomena they need help to identify. In this case, the geology crew spotted a large, reddish jellyfish that didn't seem quite like any known to biologists.

Matsumoto thought the jellyfish might be new, but proclaiming a new species is not as easy as it was in the eighteenth and nineteenth centuries. Marine biologists have to comb through the literature to make sure no one has previously described the same animal, even offhandedly, and compare the anatomy and behaviour of the mystery species to what's already documented. In this case, Matsumoto and his colleagues didn't just have a paper trail to pore over to see if anyone else had seen this jelly, but also 15 years of videos taken during deep-sea explorations.

Others had. Researchers had sighted *Tiburonia* by 1993, at least, but no one had taken particular interest in it at the time. Perhaps they didn't quite realize what they were seeing. After extensive research, Matsumoto and colleagues were able to discern that *Tiburonia granrojo* – or "big red" – really is a unique jellyfish that lives between 650 and 1,500 metres (2,132–4,921 feet) down. In fact, it's so unlike other jellyfish that the species was placed in its

own evolutionary group, the Tiburoniinae, and the name comes from the name of the remote operating vehicle the 1998 geology crew was using, the *Tiburon* (meaning "shark"). To date, the jellyfish has been seen in deep waters off California, the Hawaiian Islands and Japan. With such a broad range, the big red jelly might show up in other places, too.

If *Tiburonia* does appear on other dives around the world, marine biologists should be able to identify it pretty readily. First of all, *Tiburonia* is very large. The bell of this jelly can span 1 metre (3 feet 3 inches) across and the entire animal is a deep, vibrant red. Stranger still, the big red jellyfish doesn't have any tentacles. That's especially odd, because most jellyfish – at least those familiar to us from the seaside and aquariums – have tentacles and oral arms equipped with specialized stinging cells trailing beneath their bell. When an unlucky fish swims into the oral arms, harpoon-like stingers pop out and injure or incapacitate the fish with venom as it is drawn upwards towards the mouth. But *Tiburonia* is different.

Rather than having oral arms and stinging cells, *Tiburonia* instead possesses between four and seven flexible and fleshy arms that likely allow the jellyfish to grasp its food. What that food is, though, is as yet a mystery, but it's likely to be small morsels that are more readily available at depth.

Watching *Tiburonia* move through the water is an almost hypnotic experience. Like other jellyfish, the big red jelly primarily

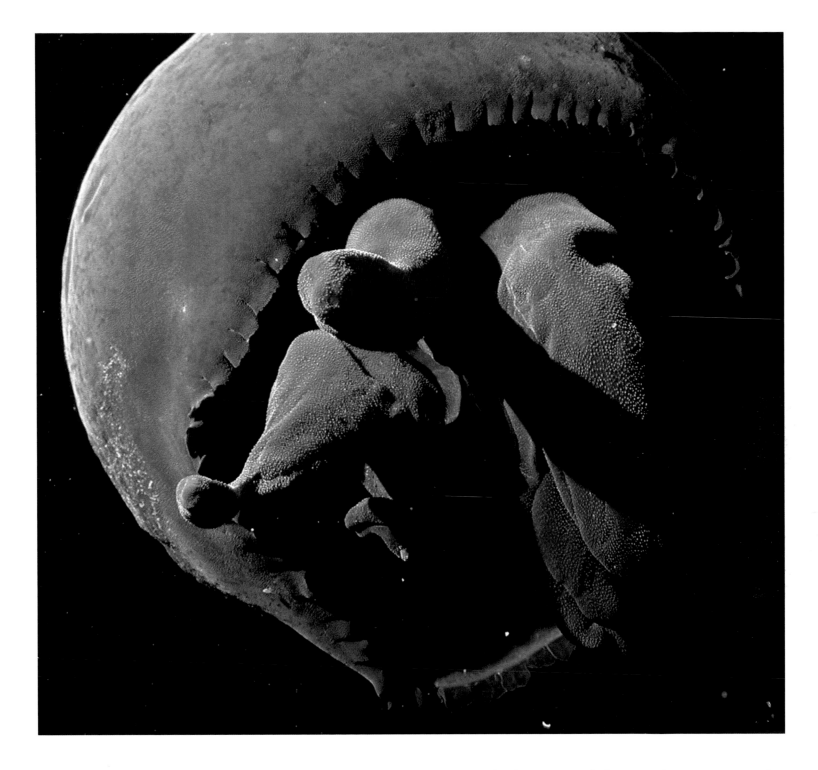

moves by flexing and relaxing rings of muscle within its large bell. The arms flex and undulate beneath, the entire animal seeming to move in slow motion.

Despite being seen multiple times and in many parts of the ocean, there's still much that remains unknown about this deep-sea giant, and only one specimen has been collected for study at the surface. But the new species was also a reminder to scientists that the deep sea is still full of organisms awaiting discovery – perhaps even some that have previously been seen but not recognized as unknown. *Tiburonia* was seen at least as far back as 1993, came to scientists' attention in 1998, and was formally named in 2003. How many other possible new species have been seen in that same time-frame but await their scientific recognition?

Above The big red jelly was seen several times by undersea explorations before it was conclusively identified as a new species.

brachiopod lived 200 million years ago, for example, then experts can hypothesize that any ancient environment in which the same species is found is about 200 million years old – handy when a rock layer might be difficult or impossible to get a direct date from.

Brachiopods were incredibly successful for about the first 280 million years of their history, during the Palaeozoic era – or 541 to 251 million years ago. Ancient reefs were sometimes built primarily by brachiopods, and many of these ancient species were filter-feeders that strained plankton and detritus from the water. But the worst mass extinction of all time cut back the brachiopods' success. Around 252 million years ago, intense volcanic activity caused the climate to rapidly warm, the oceans to acidify, and atmospheric oxygen to drop. The effects reached far into the seas and, though they survived, brachiopod numbers would never be as abundant as they once were.

Still, brachiopods have hung on even as bivalves have become much more common. You can still find them, if you know what to look for, from the shallows to the continental shelf to the deep ocean floor. To date, about 63 species of brachiopods live at depths below 2,000 metres (6,561 feet) – about a third of all living brachiopods. The deep sea is not necessarily a bad place for brachiopods to take hold. They can survive on little food, using a specialized organ called a lophophore – or a bunch of small tentacles – to circulate water through their shell and sieve out their food. They don't have to move much or find prey. Brachiopods can simply sit on or in the sea bottom, circulating the water to extract the food they need.

Deep water seems to suit brachiopods. Many of them are found in places where there is some sort of shelter from rough water or turbulent currents. Without sunlight or interference from the waves above, brachiopods can readily make a home at depth – often populating the same spot for so long that young brachiopods sometimes grow atop the shells of older ones. Perhaps it's not the most dramatic lifestyle for creatures that have been around since the dawn of animal life, but it's worked for them.

Opposite Brachiopod fossils are important touchstones for palaeontologists. They are so numerous that experts can track evolution and extinction across time by studying these ancient shells.
Above Brachiopods have existed for over 541 million years.

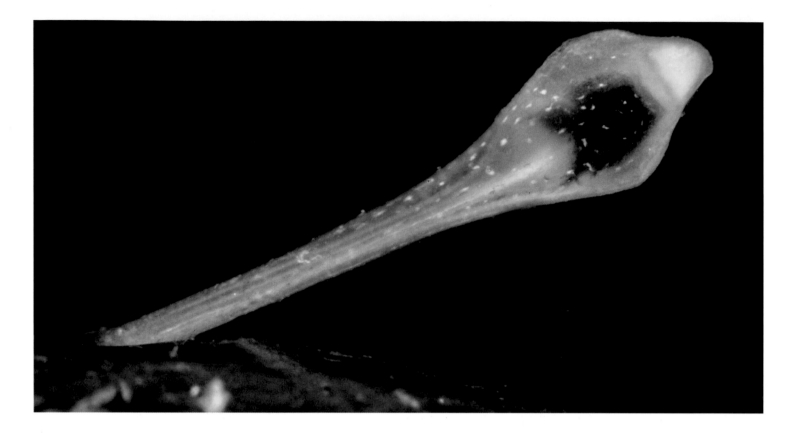

The relationship between males and females in these anglerfish has required some substantial anatomical modifications. Males often have a strong sense of smell and specialized eyes to help them locate a female. Many cannot eat – their jaws and even their throats are effectively vestigial, making it impossible for them to feed. But if a male finds a female, he attempts to bite her and attach himself. If successful, the immune system of the female anglerfish doesn't even realize that it's being attacked. The male stays latched, releasing an enzyme that erodes the female's skin and tissue to allow the male to fuse to her body. Sometimes, in fact, multiple males may fuse to a single female at once, tapping into her blood vessels to gain their life support. The males then do nothing more than hitch a free ride and are immediately available when the female is ready to spawn.

Not all anglerfish reproduce this way. Some species have females and males that are able to catch their own food and spawn more like other bony fish. But the symbiotic method used by some species speaks to the challenges of living deep, when the basics of life cannot be taken for granted.

While some big-mouthed fish – like gulper eels – use their mouths like nets to catch large quantities of small prey, anglerfish often seek out larger morsels. These fish not only possess large jaws full of prominent teeth, but their bellies can expand to accommodate prey twice as large as the anglerfish itself. Such a meal can nourish an anglerfish for a long time, a way to make the most of whatever they might find in deep water. These fish don't chase their prey, but instead try to bring the prey close enough to strike – a much more energy-efficient strategy when food is hard to come by. And if an anglerfish finds a good fishing spot, they might stay there. At least one deep-sea anglerfish has been seen hanging upside down in the water column, with its lure hovering above the burrows of small creatures. These fish not only have lures, but sometimes they go fishing themselves.

Above From the shallows to the deep sea, anglerfish are defined by their specialized lures that help entice prey closer to their jaws.

Opposite Some anglerfish species show extreme sexual dimorphism, with females being large, free-swimming predators and males living a parasitic lifestyle on the larger females. Here, two males hitch a ride.

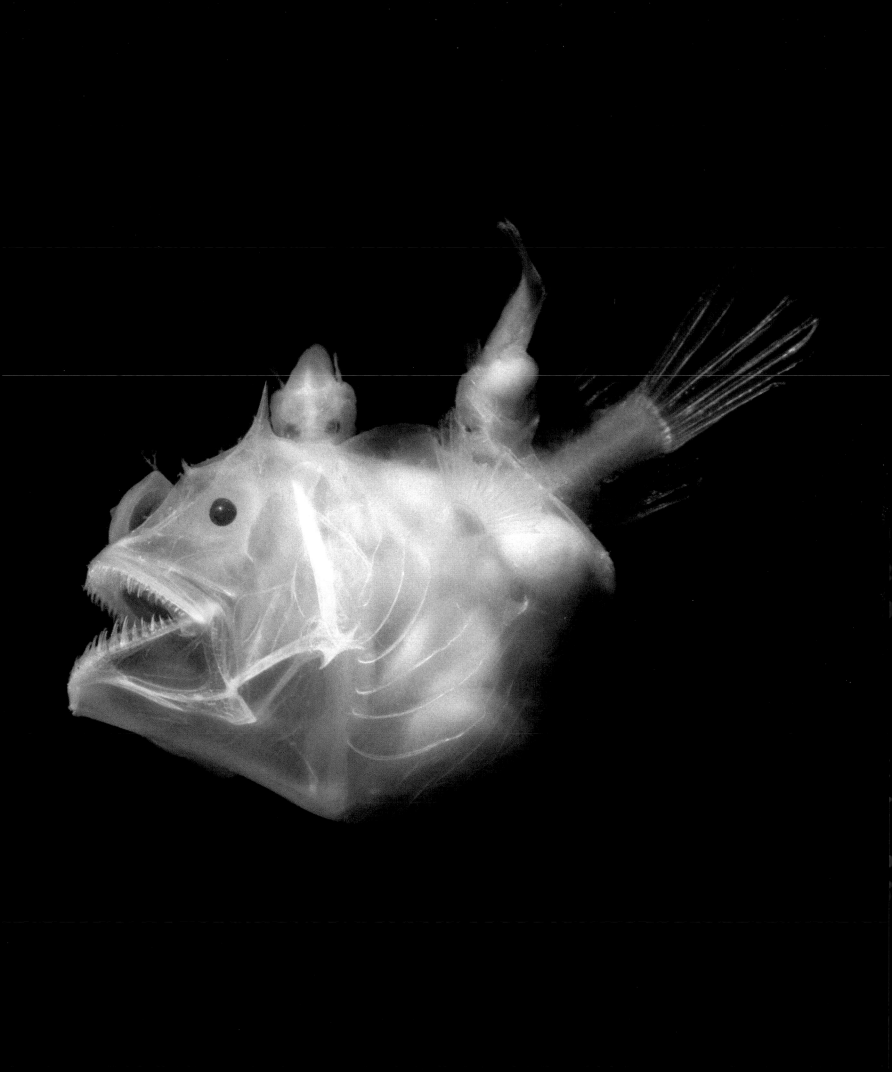

over 3,000 during the course of 12 hours. Many were of seemingly lifeless seafloor made up of relatively fresh lava. But 13 depicted an area absolutely covered in mussels, an abundance no one ever expected to see so deep down. Some of the photos also showed a strange shimmer, a possible distortion caused by high heat.

A trio of researchers quickly decided to visit the site for themselves in Deep Submergence Vehicle (DSV) *Alvin*. Nothing could have prepared them for what they saw. There were cracks in the lava field where super-hot water spewed out, turning cloudy as minerals in the heated water began to precipitate back out and fall to the seafloor. The crew found clams, crabs and octopus in the area, and subsequent dives uncovered additional sites with other forms of life – including a place they named the "Garden of Eden" that was positively covered in huge tube worms with bright red plumes (see pages 144–147). The experts had not only discovered hydrothermal vents, but that many forms of deep-sea life had adapted to make the most of these undersea hot spots.

Marine scientists had unknowingly stumbled upon traces of such vents before. Research in 1949 and 1960 found extremely hot and salty depths in the Red Sea. The records were treated as anomalies until after the vents were rediscovered in the 1970s – with many more to follow. To date, researchers have identified over 200 hydrothermal vent fields in oceans around the world. There are certainly hundreds more that have yet to be recorded.

No two hydrothermal vents are quite alike in all details, but their basic formation is the same. The vents are most often found along mid-ocean ridges where magma comes close to the seafloor. Cracks and fissures in the seafloor allow water to come in contact with rocks beneath the surface that have been heated by the magma, the heat causing the previously chilly seawater to soak up more minerals before the water is vented back out above the seafloor. As the water mixes with the cold surrounding water,

the minerals precipitate out and sometimes lead to the formation of chimney-like structures given names like "white smokers".

But the most startling aspect of hydrothermal vents couldn't immediately be detected. The higher water temperatures around the vents were not the only reason researchers found far more life than they were expecting. As experts began to study vent organisms – like the wonderful *Riftia* tube worms – they realized that these organisms had evolved an entirely different food web, one which is based upon bacteria that are able to feed on the chemical compounds spewed out by the vents. The mineral-rich water provides ample food for bacteria, particularly compounds rich in sulphur, and organisms like the tube worms are colonized by these bacteria. Many vent organisms don't need to hunt or eat food in the same way other deep-sea creatures do because they are fed by the byproducts of the chemical-hungry bacteria inside them, a process called chemosynthesis. This process is how life can thrive in a world that never sees sunlight.

Hydrothermal vents are so new to science that we are only just beginning to understand their implications. Life on Earth isn't wholly dependent on the sun, but instead can evolve alternative ways of feeding and surviving. The fact that evidence of hydrothermal vents have been found on Saturn's moon Enceladus and Jupiter's moon Europa, for example, has raised the possibility that chemosynthetic life exists elsewhere. And that may return experts to the origin of life on Earth itself. Perhaps life did not begin in the sunlit shallows, but deep in the ocean dark. The only way to know is to keep going back down.

Opposite above Mussels and shrimp at a hydrothermal vent.
Opposite below Photograph from the NOAA Submarine Ring of Fire 2006 expedition, which explored submarine volcanoes in the Pacific, shows life around an active vent (top right).

Yeti Crabs

UNTIL RECENTLY, NO ONE KNEW THAT YETI CRABS EXISTED. AMONG THE LATEST DEEP-SEA CREATURES TO BE RECOGNIZED BY SCIENCE, THESE CRUSTACEANS WERE GIVEN THE GENUS NAME *KIWA* IN 2006. BUT THESE INVERTEBRATES AREN'T LIKE THE GHOST OR FIDDLER CRABS YOU MIGHT SEE SCUTTLING ALONG THE SHORE. NOT ONLY DO THESE LONG-ARMED, BRISTLY CRABS LIVE DEEP, BUT THEY ARE PRINCIPALLY FOUND AMONG HYDROTHERMAL VENTS AND METHANE-RICH ENVIRONMENTS CALLED "COLD SEEPS".

Despite several species being identified and named so far, marine scientists didn't discover the first-known yeti crab species until 2005. Researchers aboard the DSV *Alvin* were 2,200 metres (7,217 feet) below the surface off Easter Island when experts aboard noticed the unusual crustaceans. The name "yeti crab" immediately came to mind because of the decapod's long, almost hairy-looking arms, and analysis of collected crabs confirmed that the invertebrate represented a new species, genus and even family. This first crab was named *Kiwa hirsuta*, and it didn't have to wait long to gain company. Just a year later, researchers on a dive off Costa Rica found what would eventually be named *Kiwa puravida*, with several more added to the group since those first finds.

Like many crabs, *Kiwa* are scavengers. They use their short and powerful claws to snip off parts of whatever carrion falls to the bottom. But that's not all. The arms of yeti crabs are covered in setae – bristly projections that give the crustaceans their common name. Bacteria that can break down sulphur compounds have been found on the setae of Yeti crabs, suggesting that the spines are not for protection but instead act as gardens for bacteria that

the crab can then eat in between lucky meals of decomposing organic matter. One *Kiwa puravida* crab has even been observed moving its arms in a repetitive and cyclic motion. Experts suspect this wafts mineral-laden waters towards the bacterial colonies on the crab's arms, thus feeding them and providing future snacks for the invertebrates.

Some yeti crab species look relatively similar to each other, like spinier versions of the squat lobsters found elsewhere in the deep sea. But others seem to stand apart, suggesting that there may be even more yeti crab forms waiting to be found. In 2015, marine biologists named *Kiwa tyleri*, which they nicknamed the "Hoff crab" because the setae on the underside of this species reminded them of actor David Hasselhoff's hairy chest during his time on *Baywatch*. This particular crab was found over 2,394 metres (7,854 feet) down in the southern Atlantic, the only yeti crab yet found outside the Pacific Ocean.

The fact that the "Hoff crab" has its bacteria-cultivating setae on its underside rather than its arms has puzzled biologists, which may indicate a difference in feeding strategy that's evolved over vast spans of time. Genetic studies of yeti crabs hint that they have been evolving apart from each other for millions of years. That makes sense given that hydrothermal vents open and cool over time, being temporary spots on the seafloor, and they are widely dispersed. The eggs and larvae of yeti crabs travel far and require enough good fortune to settle near a vent ecosystem in order to survive – almost like islands underwater. Different vent systems have different particulars, or at least act as somewhat isolated evolutionary pockets, and so even organisms of the same family can evolve distinct ways of tackling the same biological requirements.

The way these crabs live may have helped conceal their presence even as marine scientists became increasingly enchanted with hydrothermal vents. The crabs are relatively large, about 15 centimetres (6 inches) long, but the first expedition to find them noted that yeti crabs often hide behind or underneath rocks around the hydrothermal vents. Often, little could be seen of them beyond the tips of their arms. Once the experts knew what to look for, however, they began to notice the crabs foraging for open mussel shells and holding their hairy arms over warm water. Those distinctive arms are important for another reason. Yeti crabs don't have eyes and cannot see, meaning that the setae on their bodies have to detect the chemical signals and changes in pressure that allow crabs to navigate and forage in their undersea habitats. The crabs live in a way that is entirely alien compared to what we're familiar with on the surface, and yet they have been down there, surviving in the lightless depths of the ocean.

Below The fluffy projections along the arms of yeti crabs give these crustaceans their name.

Methanogenic Bacteria

NO ORGANISM IS AN ISLAND. EVEN A SELF-CONTAINED CELL,
LIKE A FORAM IN ITS SPIKY SHELL, RELIES ON OTHER ORGANISMS
FOR FOOD. MANY ORGANISMS RELY ON SYMBIOSIS – WHEN TWO
DIFFERENT SPECIES SHARE SPACE OR EVEN THE SAME BODY,
USUALLY TO THE ADVANTAGE OF BOTH – SUCH AS THE SPONGES
A GIANT SPIDER CRAB STICKS TO ITS SHELL OR THE *E. COLI* IN
THE GUT OF A WHALE.

Microorganisms, especially, often live within the bodies of animals and can do everything from helping digest food to creating bioluminescent lures. Researchers are still discovering such interactions in the deep ocean, including among microorganisms known as methanotrophs that survive on methane.

Microorganisms that rely on methane are key to a great deal of the deep sea's biodiversity. In addition to the hydrothermal vents, marine geologists have also found deep-sea methane seeps – fissures in the seafloor where the gas bubbles up from deeper, carbon-rich rocks. They are important sites in the Earth's carbon cycle, where microorganisms feed on carbon in rocks below the seafloor and release methane in the process. That methane, in turn, becomes food for other bacteria that then can become a food source for animals or become incorporated into the bodies of sea creatures. And much like *Riftia* (see pages 144–147) and other animals around hydrothermal vents, small tube worms of the methane seeps have found a way to make the most of bacteria that can convert natural compounds into food.

In 2020, marine scientists announced that they had found a way in which some tube worms can use methane-eating bacteria to feed themselves. In deep water off the coasts of California and Costa Rica, researchers found methane-eating bacteria in the plumes of the tube worms *Laminatubus* and *Bispira*. In fact, what drew the attention of biologists to these two species in particular is that their plumes seemed a bit fluffier and puffed out than other species. Marine scientists have begun taking this characteristic as a sign that deep-sea organisms are increasing the surface area of their bodies – much like the setae on the arms of yeti crabs (see pages 140–141) – to cultivate colonies of chemosynthetic bacteria.

Methanotrophs aren't only found in the deep sea. Wherever you can smell a natural methane source – such as from a salt marsh – there are methanotrophs present. But those that live in the dark parts of the oceans have garnered special attention for two reasons. The first is that methanotrophs that live on and in other organisms provide food to species that might otherwise not be able to exist in such hostile habitats. Just like the bacteria

around hydrothermal vents, they are an essential part of deep habitats that rely on chemosynthesis. But the second is far more important to us.

Methane is a potent greenhouse gas. When released into the air, methane is more than 25 times as effective at trapping heat in our atmosphere than carbon dioxide. If all the methane from the bottom of the sea were to bubble up and be released, the effects could be swift and disastrous. Thankfully for us, methanotrophs and the creatures they live inside alter methane and prevent the gas from reaching the surface, instead keeping it within ocean habitats. Researchers have estimated that these bacteria might trap and transform about 90 per cent of the methane seeping from the seafloor, an essential ecosystem service for all life on Earth.

The relationship between methanotrophs and the organisms they live within goes back a very long way. Even though fossils of deep-sea organisms are rare, palaeontologists have nevertheless found evidence of tube worms like *Laminatubus* and *Bispira* around prehistoric methane seeps preserved in Jurassic rock. Their presence there was not understood until the discovery of the methane-dependent worms at the bottom of our modern oceans. The mere presence of the ancient worms in the seep environment hints at a similar, if not the same, relationship between the worms

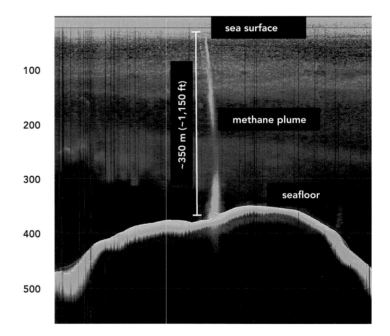

and the bacteria – an ancient case of symbiosis that has helped to protect the terrestrial world from excess greenhouse gases since the time *Stegosaurus* walked the Earth.

Top Small tube worms are among the deep-sea creatures that rely on methane-hungry bacteria.

Above Sonar image of a methane plume in the Atlantic Ocean.

143

Giant Tube Worms

IN FILM AND FICTION, WE OFTEN WONDER WHAT IT WOULD BE LIKE
TO MEET AN ALIEN SPECIES – AN ORGANISM WHOSE BIOLOGY IS SO
FUNDAMENTALLY DIFFERENT FROM OUR OWN THAT WE WOULD
BE NOTHING LESS THAN ASTONISHED. WHILE LIFE ELSEWHERE IN
THE UNIVERSE HAS YET TO BE DISCOVERED, RESEARCHERS HAVE
NEVERTHELESS FOUND EARTHLY SPECIES THAT HAVE CAUSED THEM TO
RECONSIDER WHAT THEY THOUGHT THEY KNEW ABOUT BIOLOGY ON OUR
PLANET. ONE OF THESE CREATURES IS *RIFTIA PACHYPTILA*
– GIANT TUBE WORM OF THE DEEP SEA.

Riftia are immediately recognizable. Aside from being enormous, reaching lengths of over 3 metres (9 feet 10 inches), these cousins of the average beachside sandworm have a ghostly white outer tube from which their blood-red plumes extend. They often live in great colonies, dozens and dozens of tubes growing next to and atop one another, over 2,600 metres (8,530 feet) below the surface. The worms are responsive, able to retract their red plumes into their bodies if bothered like other tube worms do, but the fact that they make permanent homes near undersea vents was a clue to just how unusual these invertebrates are.

No one had seen anything quite like *Riftia* until 1977. In that year, the crew of the DSV *Alvin* (see pages 168–173) was studying deep-sea hydrothermal vents off the Galapagos Islands when they spotted vast colonies of the worms around the openings in the ocean floor. The expedition was led by geologists – not biologists – but they took enough samples for zoologists to get a better look at these unexpected annelids (or segmented worms).

At first, the abundance of the worms around the undersea hydrothermal vents was a mystery. The deep sea is very cold, about 4°C (39°F), but the water around the vents was incredibly hot, over 350°C (662°F). These worms weren't just hanging on, they were thriving – along with crabs, clams and other invertebrates found among the nests of *Riftia*.

Dissections of the worms only revealed them to be stranger still. *Riftia*, it turns out, do not have a digestive system. How could these worms survive without one? The answer, biologists eventually discerned, is chemosynthesis.

Thanks to the outpouring of the thermal vents (see pages 136–139), the water in these deep-sea habitats is loaded with carbon, sulphur, nitrogen and oxygen. These create an abundance of carbon dioxide, sulphide, and other molecules that the worm

Opposite Enormous tube worms with bright red plumes were among the most striking early discoveries around hydrothermal vents.

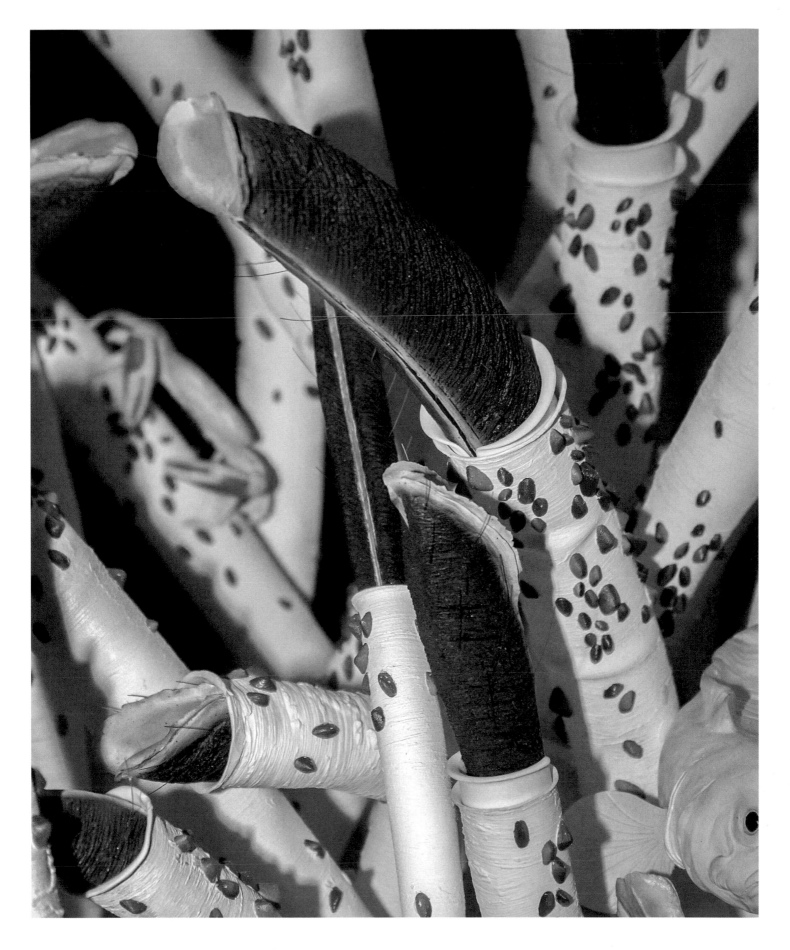

retains in its feathery, bright red plume and moves them to another part of its body called a trophosome – a vascular organ inside the worm that is absolutely brimming with bacteria. As the bacteria feed on the chemicals provided by the worm, they give off byproducts that the adult worm, in turn, feeds on.

Strange as it might seem, *Riftia* do not spend their entire lives glued to the rocks around hydrothermal vents. In fact, that would be a terrible survival strategy. Hydrothermal vents open and close over time, often with the movements of the tectonic plates that make up the Earth's crust. If a hydrothermal vent is pushed away from the underlying hot spot that created it, then the organisms that rely on the heat and outflow of the vent would entirely perish. Fortunately for *Riftia*, they start life as swimming larvae.

The entire lifespan of *Riftia* is a remarkable example of co-evolution, not just between the worm and its life-giving bacteria but between the worm and its environment. A larval *Riftia* doesn't look like much more than a small blob with little hairs called cilia to push it around. These larvae swim through the ocean depths, roving distances over 100 kilometres (62 miles) if necessary, and many never find a suitable home. But a lucky larva may settle near a hydrothermal vent and glom itself on to the sea bottom nearby, ready to change into a large, red-plumed adult.

Microorganisms around the vent are crucial for the transformation. When a larval *Riftia* finds a suitable place to settle, the worm doesn't have its essential bacteria. *Riftia* only get the bacteria they need from the surrounding environment. The bacteria colonize the inside of the worm and cause its anatomy to change in response, forming a trophosome that will provide the worm with the bacterial byproducts it needs for food. Within two years, the *Riftia* become adults and are sexually mature, creating vast mats tens of metres/feet wide of worms ready to reproduce. When the time is right, triggered by an as-yet-unknown cue, the male worms release sperm into the water that then enter the tubes of the females to fertilize their eggs. The swimming larvae soon leave to find new homes in the dark of the ocean, continuing the life cycle of this unusual annelid.

Right *Riftia* worms get their nutrition from specialized bacteria in their bodies that feed on chemical compounds in the water.

Chimaeras

WHEN WE THINK OF FISH WITH SKELETONS MADE OF CARTILAGE, SHARKS AND RAYS ARE THE MOST FAMILIAR. MANY LIVE ALONG THE COASTS AND NEAR THE SURFACE, EASY TO SPOT IN THE WATER OR WASHED UP ON THE SHORE. BUT THESE FAMOUS, FLEXIBLE FISH AREN'T THE ONLY SWIMMERS OF THEIR KIND. SHARKS AND RAYS HAVE AN ENTIRE GROUP OF DEEP-SEA RELATIVES. THEY'RE VARIOUSLY KNOWN AS SPOOKFISH AND RATFISH, AMONG OTHER NAMES, BUT MANY OF US KNOW THEM AS CHIMAERAS.

One look at a chimaera and it's easy to understand why they're sometimes called "ghost sharks". They share a lot in common with sharks – from their triangular dorsal fins to their streamlined shape – but the resemblance begins to fade from there. Modern chimaeras are almost universally small, the largest of known species being about 1.5 metres (5 feet) long. Chimaera species often have large eyes and blunted snouts that make some of them look much more adorable than their toothy shark relatives. Then again, male chimaeras have a special, grasping sexual organ called a tentaculum on their heads.

Marine biologists have been fortunate enough to observe how the tentaculum works. Odd as it may seem to have a sexual organ on the head, far from the claspers of the male or vent of the female, the placement of this structure works for these fish. During courtship, males use their tentaculum to press down and hold on to the pectoral fin of the female, locking the two fish together for the duration of their encounter.

There are a few other features that help distinguish chimaeras from sharks. While most sharks have five gills – with a few exceptions having six or seven – chimaeras tend to have four. And instead of rows and rows of teeth that pierce or stab, chimaeras have three sets of tooth plates that help them crush and grind food, similar to some rays and other bottom-feeding fish. In a sense, chimaeras are a bit of a mix of a shark body shape with the feeding preferences of a ray.

These strange fish have been swimming around the seas of our planet for a very long time. Even though the earliest fossil chimaera is around 330 million years old, biologists suspect that they evolved even earlier – around 420 million years ago. They didn't start off as deep-water fish. Many chimaeras lived in the shallows, and some were incredibly strange. *Helicoprion* was a cousin of the modern ratfish that could grow to the size of a large shark and had a circular-saw-like whorl in its lower jaw that it used to shuck ancient cephalopods out of their shells. It was only over time, as chimaeras went extinct in the upper ocean waters, that they became dedicated denizens of the deep.

Surviving in deep water often relies upon making the most of whatever happens to drift by, fall to the bottom, or live in

Opposite The enigmatic whorl-tooth fish *Helicoprion* was an ancient relative of modern ratfish.

Above Chimaeras are sometimes called "ghost sharks" for their spooky appearance.

the sand. Chimaeras use electroreception – a skill shared with sharks (see goblin shark, pages 88–91) – to pinpoint small prey such as worms, octopus and crustaceans, sucking them into their mouths and crushing the titbits with plate-like teeth. Chimaeras won't turn down carrion, either, and sometimes remote operated vehicles encounter their haunting silhouettes around whalefalls and other sources of deep-sea carrion. Some chimaeras even have specialized anatomy to help them better find and snaffle up tiny prey from the sea bottom. The Australian ghostshark lives off the coast of Australia and New Zealand, and uses a shovel-shaped, trunk-like extension of its nose to probe through the muddy bottom for food.

Dozens of chimaeras have been named by researchers, but many remain poorly known. Studying the anatomy of specimens taken from the deep or that wash up on shore is one thing, but trying to investigate the behaviour of the living animals is another. Even

so, there are some chimaeras that marine biologists have been able to discern a little more about. In the waters of the Pacific Northwest of North America, divers sometimes encountered spotted ratfish, *Hydrolagus colliei*. These chimaeras grow to about 97 centimetres (3 feet 2 inches) long, with males being smaller than females. They're immediately recognizable by a dappling of white spots against the red of their body. And though many ratfish move themselves about by flapping their pectoral fins, the spotted ratfish takes it a bit further, with underwater acrobatics that have been compared to barrel rolls. Even in the darkness of the sea, some fish show off.

Below The elephantfish is a species of chimaera that uses its peculiar snout to search for invertebrates in the sand.

Opposite Male chimaeras have a small grasping organ called a tentaculum on their foreheads that helps them hold on during mating.

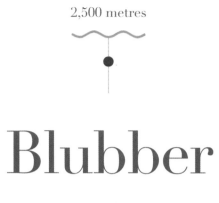

Blubber

WHERE WOULD WHALES BE WITHOUT BLUBBER? THE FATTY TISSUE
IS INSULATION, FAT STORAGE, A BUOYANCY ASSISTANT, AN ENERGY-
RICH SOURCE OF FOOD FOR PREDATORS, AND MORE, ALL HIDDEN
BEHIND THE BODY WALLS OF MANY CREATURES WHOSE ANCESTORS
TRANSITIONED FROM LIFE ON LAND TO LIFE IN THE SEAS.

Among modern animals, you can find blubber beneath the skin of whales, seals and sea lions, penguins, and manatees. That's astonishing. None of these creatures are close relatives of each other – penguins and manatees last shared a common ancestor more than 312 million years ago – and yet they have all evolved the same solution to the challenges of living at sea. In fact, not only has blubber evolved multiple times among different seagoing vertebrates, but it's far older than anyone expected.

Fish-like reptiles called ichthyosaurs proliferated through the world's oceans between 236 and 96 million years ago. Some of them, like *Ophthalmosaurus*, have often been cited as examples of convergent evolution with dolphins (see pages 62–65). But the resemblance isn't just in the bones. In 2018, palaeontologists announced that the Jurassic, 180-million-year-old ichthyosaur, *Stenopterygius* had blubber. The fossil, found in a quarry in Holzmaden, Germany, preserves portions of the marine reptile's skin. Palaeontologists have recently confirmed that such fossils often retain traces of their original biomolecules, and, in this case, the ancient skin retained some of the hallmark biochemical indicators of blubber. The fatty tissue likely helped

These pages Many different organisms have independently evolved blubber, including whales, penguins, seals and sea lions, and even ichthyosaurs.

the ichthyosaur maintain a warm body temperature – estimated to be about 35°C (95°F) – especially if the reptile ventured below the sunny surface waters.

Instead of evolving just once, blubber has originated time and again among groups of land-dwelling vertebrates whose descendants made a home in the seas. In most cases, the lipid-rich tissue covers almost the entirety of an animal's body, except the limbs, and is connected to the underlying structure of the body with tendons and ligaments. And some animals really pack it on. Depending on time of year and age, for example, blubber might make up fully half of a whale's body mass.

Understanding what blubber does helps to explain why this very useful tissue has evolved time and again. For one thing, marine mammals and other creatures wrapped in blubber – like penguins – often rely on the fat's ability to store energy for their survival. Mother elephant seals, for example, often build up as much blubber as they can before the pupping season. They have to protect and feed their offspring while on shore for 8–10 weeks, unable to return to the sea to eat without risking the survival of their babies. Instead, the mother elephant seals subsist by burning off their blubber, a source of energy, but also of hydration, as metabolizing the fats also releases enough water to prevent the seals from drying out.

And for diving deep down, where temperatures plummet and water pressure increases with every metre/foot of depth, there could hardly be a better adaptation than blubber. While some diving animals keep warm with fur or feathers, the problem with these insulating coats is that they trap air, which can then be squeezed out with increasing depth. Blubber, being squishy and internal, can compress with pressure, and as an added bonus, the blood vessels around the blubber can squeeze shut to keep all the warm blood circulating at the animal's core. Blubber is part of what allows cetaceans like Cuvier's beaked whales (see pages 156–159) to make their astonishing dives into the Twilight Zone.

Blubber also helps whales and other creatures to save energy. Part of the reason why a dolphin – or an ichthyosaur, for that matter – seems so streamlined is because the blubber beneath their skin helps to give them a smoothed, sleek form that water can more easily pass over. On top of that, fat is a buoyant tissue. It can help animals with large blubber stores stay near the surface to feed, for example, or reduce the amount of energy required for a deep-diving animal to return to the surface.

This is not to say that all creatures with blubber use it in the exact same way. Pinnipeds – seals, sea lions and walruses – usually have both fur and blubber, which take on different roles. Fur seals have a greater density of hairs on their pelts than many of their relatives, for example, and so their fur insulates them and their blubber mostly acts as energy storage. Most other pinnipeds, by contrast, tend to have more blubber and rely on the fats for both insulation and energy.

Below For walruses, blubber also acts as energy storage.
Opposite Even in warmer waters, blubber can assist with buoyancy underwater.

Cuvier's Beaked Whale

HOW LONG CAN YOU HOLD YOUR BREATH UNDERWATER? FOR MOST PEOPLE, THE ANSWER IS SOMEWHERE BETWEEN 30 SECONDS AND 2 MINUTES. WITH A LOT OF TRAINING AND SOME ASSISTANCE FROM BREATHING PURE OXYGEN, SOME PEOPLE HAVE BEEN ABLE TO STRETCH THAT MEAGRE AMOUNT OF TIME TO 24 MINUTES. BUT THAT'S STILL NOTHING COMPARED TO CUVIER'S BEAKED WHALE, KNOWN TO SCIENTISTS AS *ZIPHIUS CAVIROSTRIS*. THESE SNOUTY, MEDIUM-SIZED WHALES CAN STAY BELOW FOR 3 HOURS AND 42 MINUTES, LONG ENOUGH TO SLIDE INTO THE OCEANS' MIDNIGHT ZONE AROUND 3,000 METRES (9,842 FEET) BELOW THE SURFACE.

Even though Cuvier's beaked whales aren't nearly as famous as the great baleen whales or playful dolphins, they are still a common and widespread whale species. These whales are found in all but the coldest of ocean waters worldwide and don't look quite like most other familiar cetaceans. They have almost porpoise-like faces on long, tapering bodies topped with a small dorsal fin near the base of the tail that gives them something of a long cigar shape.

Even though Cuvier's beaked whales are still swimming through the seas today, the first scientist to describe them thought they were extinct. In 1823, the French anatomist Georges Cuvier wrote about an unusual whale skull found on the Mediterranean coast that didn't match any recognized species. Given that Cuvier believed that much of the world had already been well explored and most large animals had been documented, he assumed that the skull must be from a long-extinct cetacean species and gave the whale its scientific name. It wasn't until 1850 that another anatomist found a skull from the same species on a beach and realized that Cuvier's beaked whale was still very much alive.

Opposite Cuvier's beaked whales are named after the French anatomist Georges Cuvier.

Top Skull of a Cuvier's beaked whale. While they are technically "toothed whales", they only have vestigial teeth along the jaw.

Above Cuvier's beaked whales often feed on squid. One individual had more than 200 squid in its stomach when it died.

Among whales, Cuvier's beaked whale is what's called an odontocete – a toothed whale like dolphins, porpoises and orcas. But the name is a little misleading in this case. Both males and females of Cuvier's beaked whale have a set of small teeth that are thought to be vestigial, slowly evolving away, while the males have a set of tusks on their lower jaws.

Rather than impaling fish and other small prey on their teeth like some other toothed whales do, Cuvier's beaked whale feeds in an entirely different way. As the whale approaches a morsel – let's say a small squid – the mammal opens its jaws and sucks its tongue back to create a small vortex that draws the prey into the whale's mouth. This is called suction feeding, and it's a strategy that Cuvier's beaked whales use way down in the dark parts of the ocean, where light entirely fades out.

So far, Cuvier's beaked whale holds the record for the deepest-diving mammal. The record-setting dive was recorded off the California coast in 2014 when a whale being tracked by scientists dived to about 3,000 metres (9,842 feet). That's over 700 metres (2,296 feet) deeper than sperm whales. But that's not the only way that researchers know that these whales reach incredible depths for an air-breathing mammal. They also look at the stomach contents of stranded whales.

What Cuvier's beaked whales eat varies from place to place. The whales off the California coast feed on different prey than those in the Mediterranean, and, naturally, a deep-diving whale will have to choose among what it finds on any particular deep-water foray. In general, though, Cuvier's beaked whales appear to prefer

squid from the Twilight Zone. The whales will eat fish, shrimp, crustaceans, and other food, but squid are clearly a favourite. One particular whale from the California coast was found with 200 squid in its stomach, food likely found by echolocation as the whale navigated in the dark.

Precisely what allows Cuvier's beaked whales to break diving records isn't fully understood. The whales can dive much faster than scientists can, even in specialized equipment, and are notoriously hard to observe as they descend. One hypothesis is that the whales are able to collapse their rib cages and lungs to cope with the intense pressure as they hold their breath. Another possibility is that the whales have a slow metabolism which allows them to ease the build-up of lactic acid as oxygen is used up on a dive.

Food isn't the only reason that these impressive cetaceans go deep, however. The longest dive for a Cuvier's beaked whale yet recorded is about 222 minutes, an incredible feat. But the whale was probably not seeking out food. Less than a month before the record dive, the individual whale had suffered exposure to intense sonar signals from the US Navy. The whale might have been diving deep to avoid the noise again, a reminder that what we do near the surface can have profound consequences.

Below Stranded Cuvier's beaked whale, La Jolla, California, 1959.
Opposite Cuvier's beaked whales are the deepest diving mammals on the planet, reaching over 3,000 metres (9,842 feet) below the surface on a single breath of air.

Paleodictyon

THE DEEP SEA IS FULL OF MYSTERIES. THERE ARE SPECIES RESEARCHERS KNOW LITTLE ABOUT, CREATURES THAT HAVE YET TO BE GLIMPSED, AND LONGSTANDING ENIGMAS THAT CONTINUE TO EVADE ATTEMPTS TO SOLVE THEM. AMONG THESE PUZZLES IS A CREATURE – IF IT EVEN *IS* A CREATURE – WHICH RESEARCHERS CALL *PALEODICTYON NODOSUM*.

In 1850, the Italian naturalist Giuseppe Giovanni Antonio Meneghini described an odd, honeycomb-like fossil. The polygons in the rock were what palaeontologists would later identify as trace fossils – indentations in sediment that record the behaviour of prehistoric life, such as dinosaur tracks. These are different from body fossils, which record the form of ancient organisms themselves. But what might have seemed like just another fossil curiosity has gradually turned into a persistent conundrum that continues to nag at experts hoping to classify what *Paleodictyon* is and understand what sort of organism creates the structure.

Paleodictyon fossils are among the earliest known traces in the fossil record, dating back 541 million years – and perhaps even more – into the Cambrian period. This was a time when animal life was in its infancy on Earth, a time when many of the earliest creatures looked strange and entirely unconventional to our modern eyes. And if *Paleodictyon* was just another odd fossil, it would belong to the ranks of many other fossil oddities that are difficult to understand from across such a broad span of time. But *Paleodictyon* traces are not just found in Cambrian rocks. They have a fossil record that stretches across hundreds of millions of years. The youngest *Paleodictyon* fossils are about 50 million years old, an incredible range for any organism. But that's not all.

For over a century, *Paleodictyon* was thought to be extinct. The trace only appeared in the fossil record. But in 1978, oceanographers Peter Rona and George Merrill published a report on *Paleodictyon* sighted on the modern sea bottom. These traces were not exposed fossils, but had been freshly made. Even now, in the deep sea, something is making the characteristic honeycombed patterns of *Paleodictyon*.

Opposite Fossils of *Paleodictyon* show a distinctive, honeycomb pattern.
Above Map showing the Mid-Atlantic Ridge, where samples of *Paleodictyon* were taken in 2003.

Palaeontologists have been striving for decades to understand what *Paleodictyon* represents. The entire fossil is made up of hexagon or hexagon-like divots separated by small ridges only 1–2 centimetres (up to ¾ inch) across. This makes up a kind of mesh that looks almost like scales on patches of the sea bottom. Some experts have proposed that these structures are burrows created by small, as-yet-unknown creatures. Then again, these structures might represent the pathways of foraging creatures, a way to trap food, or traces left by the growth of an unidentified species that lives in the sediment on the ocean bottom.

At present, it seems that *Paleodictyon* probably does not represent a burrow or is in any way excavated from the sediment. It's more likely that the fossil is either some kind of imprint created by a living thing – or that these structures are made by some non-living geologic phenomenon that has yet to be recorded. The fact that *Paleodictyon* is still found in ocean habitats about 3,500 metres (11,482 feet) down makes the case all the more frustrating. This prehistoric enigma, which has been around for over half a billion years, is still here, yet no one has been able to observe how these structures are made.

That's not for lack of trying. During a 2003 expedition along the Mid-Atlantic Ridge to investigate hydrothermal vents, researchers aboard the DSV *Alvin* (see pages 168–172) found and took samples from *Paleodictyon* on the sea bottom. The hope was that the scientists might be able to capture the organism that creates the burrows, or some sort of evidence – be it body part or organic byproduct – that would offer a clue. But no sign was found of the trace's creator. This led experts to propose that whatever is creating *Paleodictyon* made the mesh to trap bacteria the creature could later feed on, but this idea has yet to be verified.

The slow pace life can take in the deep ocean might confound attempts to find and identify what *Paleodictyon* truly is. Perhaps, researchers speculate, the mesh represents the structure of a sponge or similar creature that has perished and been slowly eaten away by bacteria to only leave an outline behind. Organisms called xenophyophores – essentially armoured amoebas that are known to live in the deep (see pages 164–167) – are also candidates. Despite appearing fresh, there's really no telling how old any given *Paleodictyon* "burrow" is. With how little the deep seabed gets disturbed, such traces might last for centuries. It's like chasing a ghost.

Right The oldest *Paleodictyon* fossils are about 541 million years old, but no one yet knows what these traces really represent.

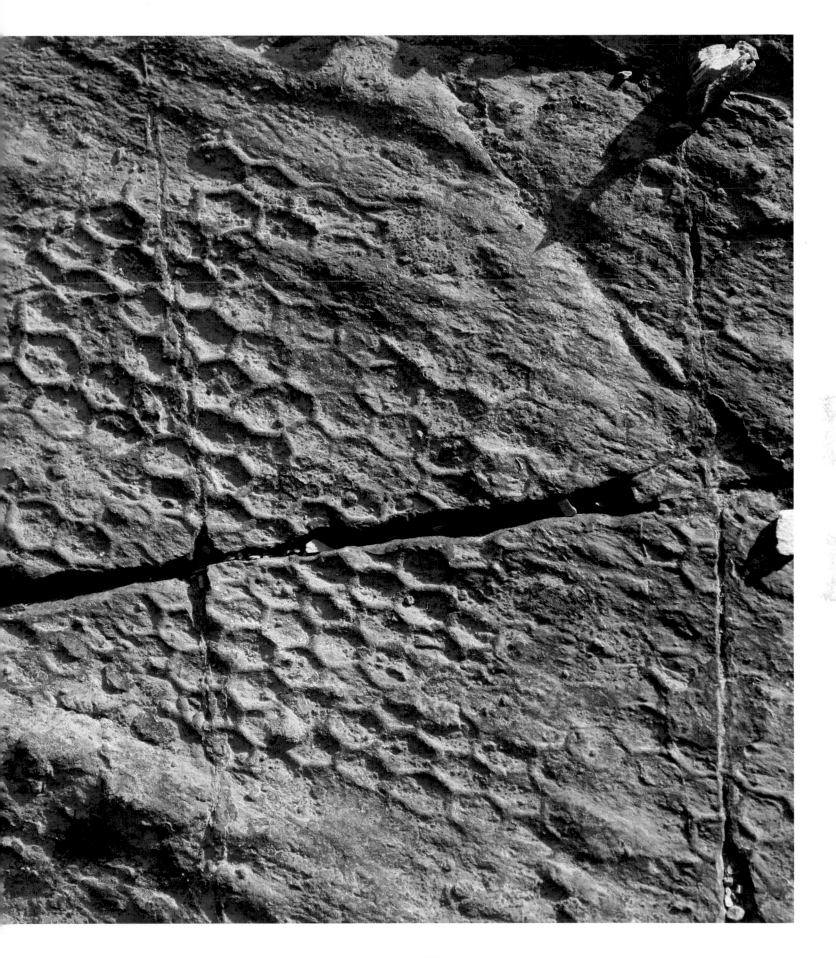

Foraminiferans

AROUND 55 MILLION YEARS AGO, THE DEEP SEA SUFFERED AN
EXTINCTION EVENT. AT THAT TIME, WHICH PALAEONTOLOGISTS DESCRIBE
AS THE PALAEOCENE–EOCENE THERMAL MAXIMUM, VAST AMOUNTS OF
GREENHOUSE GASES WERE RELEASED BY VOLCANOES AND DEEP-SEA
METHANE POCKETS. THE INFLUX OF THESE GASES CAUSED A SHARP
WARMING SPIKE THAT DRAMATICALLY ALTERED LIFE ON EARTH. ON LAND,
HUMID FORESTS BECAME DRIER, AND MAMMALS BECAME SMALLER DUE
TO ENVIRONMENTAL DISTURBANCE.

But the hot temperatures didn't just affect life on land. The warmer climate was so intense that it altered the nature of ocean currents, causing warmer waters near the surface to be shunted deeper down. The effect shocked some of the creatures that lived there, most of all a curious group called benthic foraminifera. These deep-dwelling microorganisms were like amoebas inside shells made of calcium carbonate, and about 37 per cent of their species disappeared from the deep in short order.

The foraminiferans – or forams – did not disappear entirely. In fact, they are among the hardiest and most resilient of organisms in the seas. The earliest forams evolved more than 650 million years ago, long before the first recognizable animals, and they can still be found in habitats ranging from the surface to the deepest oceans. Scientists have recognized over 4,000 living species, and there are almost certainly many more.

While not all forams have them, the most distinctive part of many forams is their shell – technically called a "test". The test can have one or several chambers and, depending on the species, can be made of different materials. But the test is not an external home, like a hermit crab's shell. The test grows within the cell membrane of the foram, enclosed within the animal's body. It's like an internal skeleton, with the chambers housing the nucleus, and organelles like mitochondria. The test also has holes that allow the pseudopods – or "false feet", much like the extensions of an amoeba – to extend out into the water and move the foram along or gather food. Forams have different ways of obtaining nutrients from filter-feeding to penetrating the tests of other forams, making this group of ancient organisms incredibly adaptable.

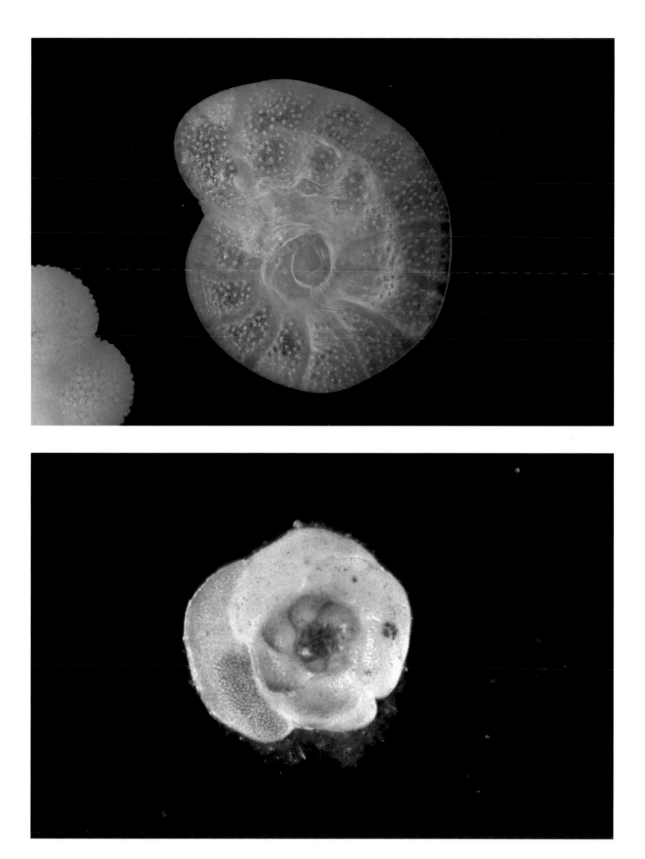

This page Foraminiferans are amoeba-like organisms with a hard shell inside their bodies, and they are found from the surface of the seas to the deepest trenches.

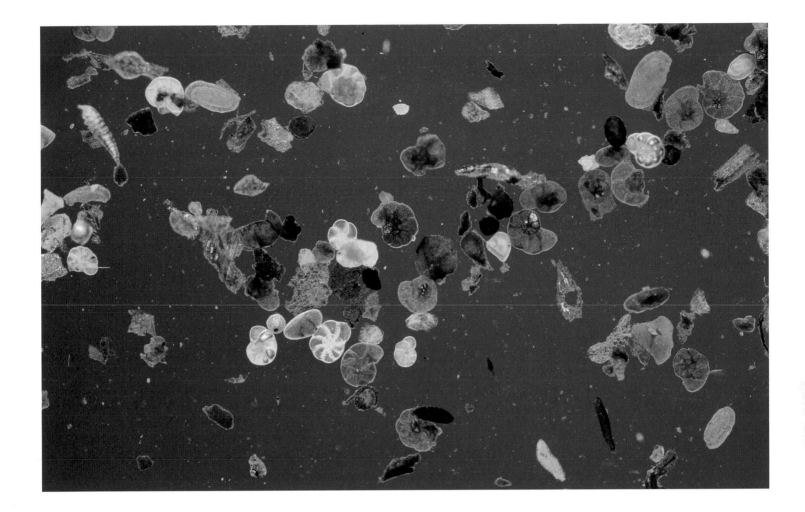

A few foram species, about 40, float as part of the oceans' plankton. Most, however, are tucked between grains of sediment on the sea bottom and pull themselves around with their thread-like appendages. They can be so numerous that sometimes the ocean sediment itself is primarily made up of their shells, the basis for the calcareous ooze that can eventually become vast accumulations of chalk, like the famous Kansas chalk of the American Midwest that often preserves the bones of ancient marine reptiles.

Despite their small size and the fact that they almost never get the spotlight, forams are an essential part of the deep ocean. A wide selection of oceanic organisms – from sand dollars to fish – eat forams. The small morsels are incredibly numerous, and that abundance has assisted palaeontologists in their efforts to study the deep past. Foram species are so distinctive and evolved so quickly that some of them can be used as a kind of biological marker tied to particular time periods, and changes in which species are present can document the speed of mass extinctions. Part of why we know the mass extinction that struck the Earth 66 million years ago was caused by an asteroid and occurred

rapidly isn't because of the land-lubbing dinosaurs, but because the fossil record of forams show a sharp and abrupt response to the impact's aftermath.

But that's just one application of the deep history of forams. Forams that live as plankton versus those that live on the sea bottom are often anatomically different and can be distinguished from each other even in the fossil record. When geologists drill ocean cores, forams can act as proxies for the depth of a particular rock layer – with more benthic species indicating a deeper habitat – and even how salty the water was at a particular site. The fact that forams have existed for so long – and, with their shells constantly falling to the sea bottom, have created what may be the best and most continuous fossil record of any group of organisms – means that the fleeting lives of today's forams are tied to a history spanning hundreds of millions of years.

Opposite Foraminiferans can move themselves around and capture prey by using arm-like projections called pseudopods.

Above After death, foraminiferans sink down through the water column and can accumulate into biogenic sediments on the sea bottom.

DSV *Alvin*

THE OCEAN IS HOSTILE TO HUMANS. WE LACK THE FINS, THE GILLS, THE BLUBBER, AND OTHER ADAPTATIONS THAT SEAGOING CREATURES RELY ON TO NAVIGATE THEIR WORLD. THE DEEP SEA IS INHOSPITABLE. EVEN WITH THE HELP OF SCUBA GEAR, THERE IS ONLY SO FAR WE CAN DESCEND BEFORE THE COLD, THE DARK AND THE PRESSURE PROVE TO BE TOO MUCH. SO GLIMPSING THE DEEP SEA HAS REQUIRED INVENTIONS AND VEHICLES THAT CAN WITHSTAND THE EXTREMES OTHER ORGANISMS SURVIVE IN, AND FEW ARE AS CELEBRATED AS DEEP SUBMERGENCE VEHICLE (DSV) *ALVIN*.

To date, *Alvin* has made more than 5,000 dives and collected data from a range of undersea environments – from hydrothermal vents to the Mid-Atlantic Ridge (see below) – that have been incorporated into thousands of scientific papers. Since the time of its first descent in 1965, *Alvin* has had an incredible undersea career.

During the time that *Alvin* was commissioned in the mid-twentieth century, researchers who were curious about the deep sea were facing a problem. Undersea vehicles called bathyscaphes – such as the *Trieste* (see pages 208–211) – had allowed scientific teams to reach incredible depths. But bathyscaphes were not very manoeuvrable underwater. They could go up and down, but they

Opposite *Alvin* on its first deep dive, 20 July 1965.

Above *Alvin* on the deck of a support vessel. The mechanical arm folded
on the front was used to gather specimens from the ocean floor.

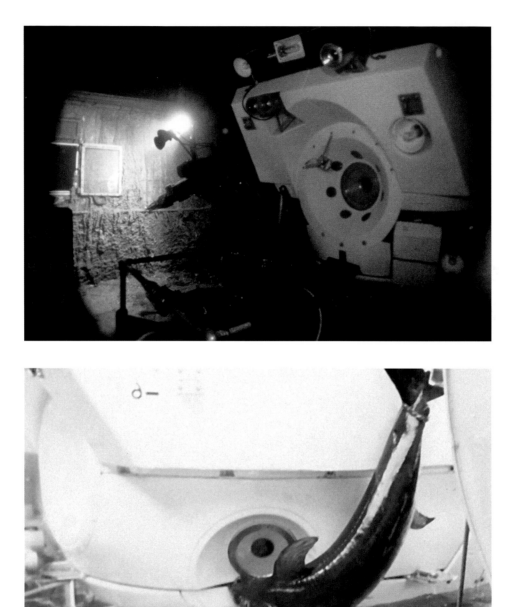

weren't especially suited to having a look around or exploring the initial area of the descent. *Alvin*, by contrast, was designed by General Mills' electronics division to be able to wander and also collect specimens with the help of two robotic arms.

Commissioned for use by the US Navy and the Woods Hole Oceanographic Institution in Massachusetts, *Alvin* is crewed by a pilot and up to two scientists. Launched from the support ship R/V *Atlantis*, the submersible can descend to 4,500 metres (14,763 feet) – just into the Abyssal Zone – during a nine-hour dive. *Alvin* was also designed with a quick-escape option in case the crew runs into trouble at depth. In the case of an emergency, the titanium sphere containing the crew can be released to float to the surface while the outer shell of the vessel can fall away for later retrieval.

Soon after its construction and certification dive, *Alvin* was put to work. In 1966, the US Navy deployed *Alvin* off southern Spain to retrieve an unexploded hydrogen bomb that had been lost when the plane carrying it collided with its fuelling tanker. *Alvin* located the bomb 910 metres (2,985 feet) below the surface and brought it back up.

But not all *Alvin*'s early forays into the deep went so smoothly. On 6 July 1967, the submersible was 610 metres (2,001 feet) below the surface when the it was attacked by a swordfish. The fish's irritation was so great that it accidentally became caught in the *Alvin*'s outer shell and caused the crew to abandon the dive. They consoled themselves by cooking the swordfish for supper. Nor was that the last mishap. A little more than a year later, *Alvin* was being lowered into the water from its support ship when two cables snapped. The crew hatch was still open when this happened, leaving the crew little time to escape before *Alvin* plummeted 1,500 metres (4,921 feet) to the bottom and couldn't be recovered until June of 1969. Despite the expensive and frustrating mishap,

Alvin was given an update and a refit to be able to descend to greater depths and was soon diving even deeper.

This extended range allowed researchers to take the submersible to places humans had never seen. Scientists curious about the Mid-Atlantic Ridge – where new rock pours out of the ocean floor to spread the adjoining North America Plate from the Eurasian and African Plates – used *Alvin* to dive there in 1974. It was *Alvin*'s crew who first discovered hydrothermal vents, and the strange lifeforms that live around them, in 1977. And *Alvin* was the vehicle of choice to visit the wreck of the RMS *Titanic* in 1986.

Yet it would be a mistake to say that *Alvin* is the same vessel today as it was in 1967 or even 1986. Every three to five years, *Alvin* was completely taken apart so each piece could be inspected, replaced or updated. Even though the deep submergence vehicle retains the same name, every single part has been replaced or changed out at least once – a constantly shifting submersible that's had its own evolution as technology and scientific needs have progressed. The last major update was in 2014 – which changed out the titanium sphere surrounding the crew while also adding new cameras and lights – and *Alvin* was quickly put back into service to assess the damage of the 2010 Deepwater Horizon oil spill in the Gulf of Mexico.

Even all these years later, *Alvin* is still making dives. Who knows what the storied submersible will find on its next foray beneath the surface?

Opposite above *Alvin* exploring the deck of RMS *Titanic*, 1986.

Opposite below In July 1967, a swordfish became caught on *Alvin* during a dive.

Following pages Refitted and updated every few years, *Alvin* continues to ferry researchers into the deep.

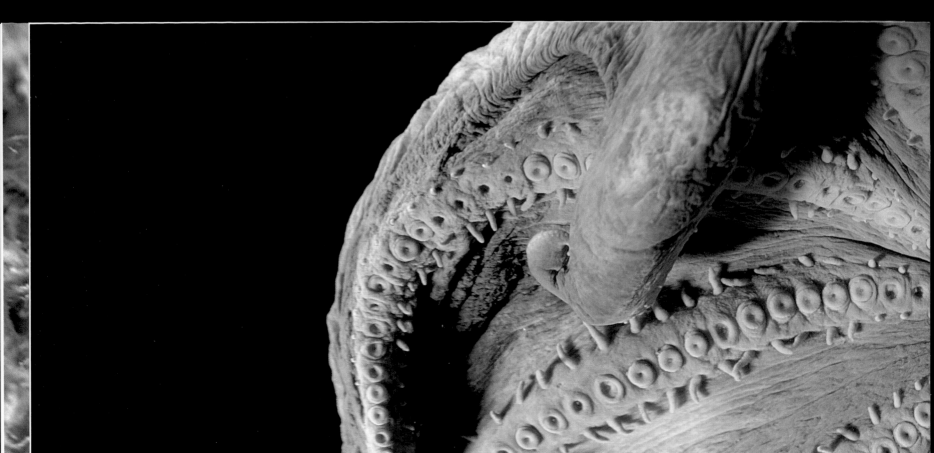

8,183 metres

HMS *Challenger*

IN THE ENTIRE HISTORY OF DEEP-SEA EXPLORATION, THERE'S ONE NAME THAT BOBS TO THE SURFACE AGAIN AND AGAIN: HMS *CHALLENGER*. EVEN THOUGH NATURALISTS HAD BEEN FASCINATED WITH THE OCEAN AND ITS DEEPEST REACHES FOR CENTURIES PRIOR TO THE SHIP'S 1872–76 EXPEDITION, IT'S THE VOYAGE OF THE *CHALLENGER* THAT IS OFTEN CITED AS THE BIRTH OF OCEANOGRAPHY. THE SHIP'S NAME IS EVEN IMMORTALIZED ON MAPS OF THE SEAS – THE MOST DISTANT PART OF THE MARIANA TRENCH, WHICH *CHALLENGER* DISCOVERED, IS CALLED THE CHALLENGER DEEP.

Challenger didn't start as a scientific vessel. Built in Woolwich Dockyard and launched early in 1858, HMS *Challenger* was a Royal Navy corvette that could use steam power when necessary. Her early history was one of colonialism; the warship was used in the occupation of Mexico's port at Veracruz and shelling of a village in Fiji. The change in the ship's story, which would eventually overshadow the vessel's wartime activities, would come at the prompting of English marine scientist Charles Wyville Thomson.

In 1870, Thomson approached the Royal Society of London with an idea for a grand expedition. Most of what people knew about the seas came from the shoreline or occasional observations by ships at sea. Many saw the oceans as incredibly sparse, almost like aquatic deserts. Yet no one had really studied the seas in detail. Thomson wanted to conduct such studies, across oceans over a multi-year voyage. And when his request was approved by

the British Government, HMS *Challenger* – stripped of guns and refitted to make space for scientific study – was the vessel of choice. The crew would include six scientists among its full complement of 200. The ship was ready to sail on 7 December 1872.

While other expeditions – such as that of the famous HMS *Beagle* that took a young Charles Darwin to the Galapagos – were imperial expeditions that carried naturalists on board, the voyage of *Challenger* was specifically scientific. Initially headed towards the Canary Islands in the eastern Atlantic, the ship covered over 127,600 kilometres (79,286 miles) and logged 362 stops during its world cruise. These breaks from the open sea were necessary

Opposite HMS *Challenger* was launched in 1858 as a war vessel and later refitted as a scientific ship.

Above Group of officers on the deck of HMS *Challenger* during the scientific expedition.

(not just for refitting). Thomson and the other scientists aboard wanted to keep regular intervals, taking depth soundings and making collections on a rhythm so that the researchers could track sea changes across the latitudes and longitudes. Whatever the naturalists brought aboard, they studied in the ship's lab , which was equipped with microscopes, specimen jars, hooks to hang larger animals, a library and all sorts of preservative equipment to make sure as much as possible made it back to England.

The amount of oceanic information the crew collected is almost incalculable. No one had examined the seas like this before. But the observation that would make *Challenger* world-famous was a stop the ship made three years into its journey. In 1875, *Challenger* was cutting its way around the southern Pacific when the crew ran into a problem. The plan was to land at Guam, but intense winds prevented the ship's landfall. So they kept going. Adjusting the route, *Challenger* stopped at the 225th sample collecting spot on 23 March, at a place between Guam and Palau. There, the crew dropped a weighted rope to the bottom – and it kept going. The depth sounding was recorded as 4,475 fathoms, or 8,183 metres (26,847 feet). No one had contemplated that the ocean could be so deep.

Challenger did more than just detect the bottom of the world's deepest trench. The crew also dredged what they could from the depths. Among the finds were fossilized teeth from one of the largest sharks that ever lived, *Otodus megalodon*, although the scientists aboard were just as excited about the ooze from the bottom of the sea. One of the sailors aboard wrote: "The mud! Ye gods, imagine a cart full of whitish mud, filled with minutest shells, poured all wet and sticky and slimy on to some clean planks." While such practices might have been anathema to seasoned sailors hoping to keep the ship clean, such samples were scientific wonderments that no one had ever seen before. *Challenger* showed how little we knew about our own planet, beginning an enduring fascination with what lies far below.

Right The *Challenger* expedition made one of the first dedicated collections of deep-sea samples: from left to right, morid cod, juvenile female anglerfish, and sediment from the ocean floor.

Following pages Track chart of the *Challenger* expedition, which covered vast distances and yet still only visited a small portion of the world's oceans.

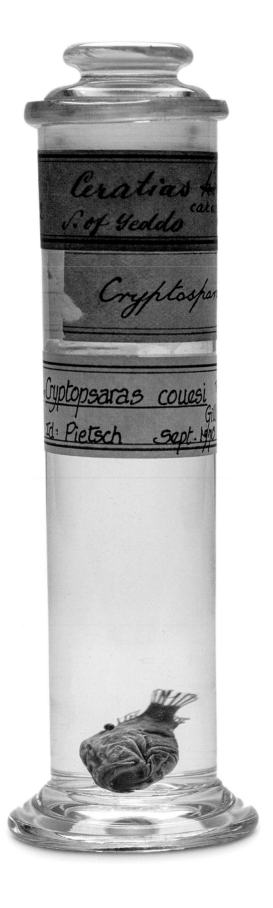

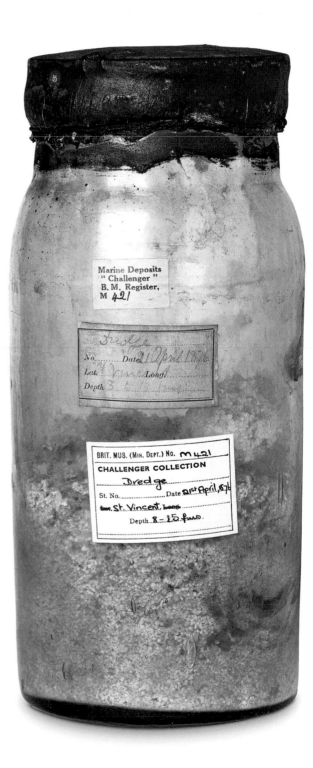

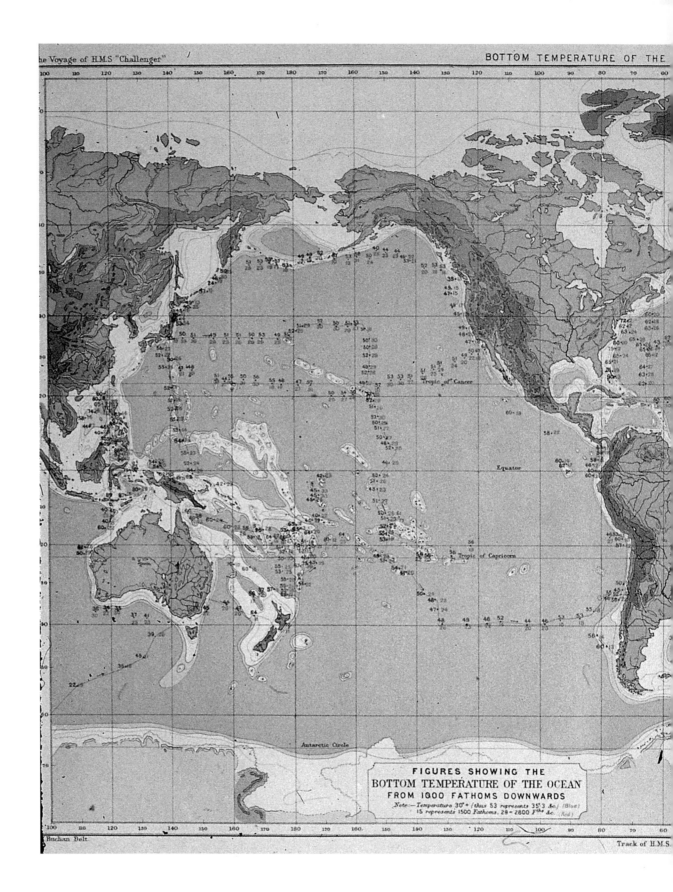

FIGURES SHOWING THE
BOTTOM TEMPERATURE OF THE OCEAN
FROM 1000 FATHOMS DOWNWARDS

Note:—Temperature 30°+ (thus 53 represents 35°.3 &c.) (Blue)
15 represents 1500 Fathoms, 28 = 2800 F.ths &c. (Red)

Buchan Delt.

Track of H.M.S.

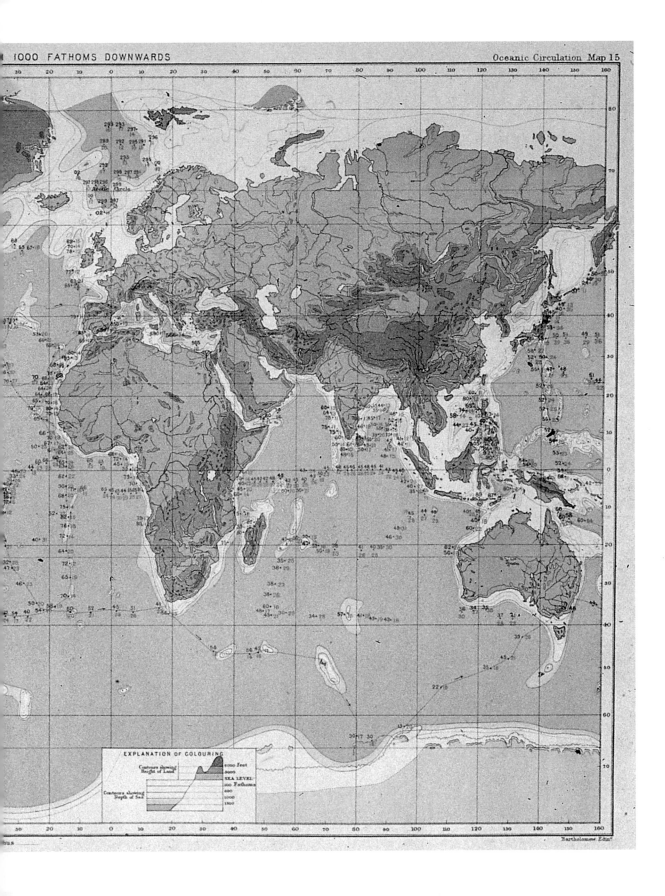

EXPLANATION OF COLOURING

Contours showing
Height of Land

6000 feet
3000
SEA LEVEL
100 Fathoms
400
1000
1500

Contours showing
Depth of Sea

Bartholomew Edin.

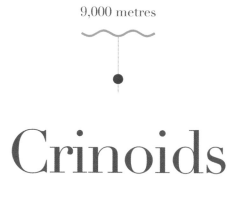

Crinoids

ANIMALS DON'T ALWAYS LOOK LIKE WHAT WE EXPECT. SQUID, FISH, CRUSTACEANS AND MANY OTHER DEEP-SEA SPECIES ARE IMMEDIATELY RECOGNIZABLE AS ANIMALS, BUT NOT SO WITH CRINOIDS. THESE VERY ANCIENT ORGANISMS ARE OFTEN ANCHORED TO THE BOTTOM, WITH FEATHERY PROJECTIONS RADIATING OUT FROM THE END OF A LONG STALK. THEY, TOO, ARE ANIMALS, AND THEY ARE SOME OF EVOLUTION'S GREATEST SURVIVORS.

Crinoids are so inconspicuous that sometimes it's easy to forget that they can still be found in modern seas. The fossil record is absolutely full of varied crinoid species that thrived for hundreds of millions of years, persisting even as more charismatic animals like trilobites and ammonites entirely disappeared. Along with brachiopods, crinoids are the unsung stalwarts of the deep sea – almost all the way to the darkest and most distant depths.

In the vast tree of life, crinoids are echinoderms – relatives of starfish and sand dollars. Mature crinoids have a mouth at the centre of their feather-like arms that sift plankton and other organic matter from the water. Once snagged, those tiny pieces of food are passed along the arms by hair-like cilia arranged along the appendages. The crinoid body is primarily centred around a cup-like arrangement of plates that hold the animal's vital organs. Most crinoids become anchored to the bottom, although there are some that swim freely in the water column or can even crawl along the sea bottom using their arms to drag themselves along. And while they might seem like part of the oceans' backdrop, they have maintained a diverse array of species despite some major setbacks during Earth's Big Five mass extinctions. Scientists have recognized about 600 crinoid

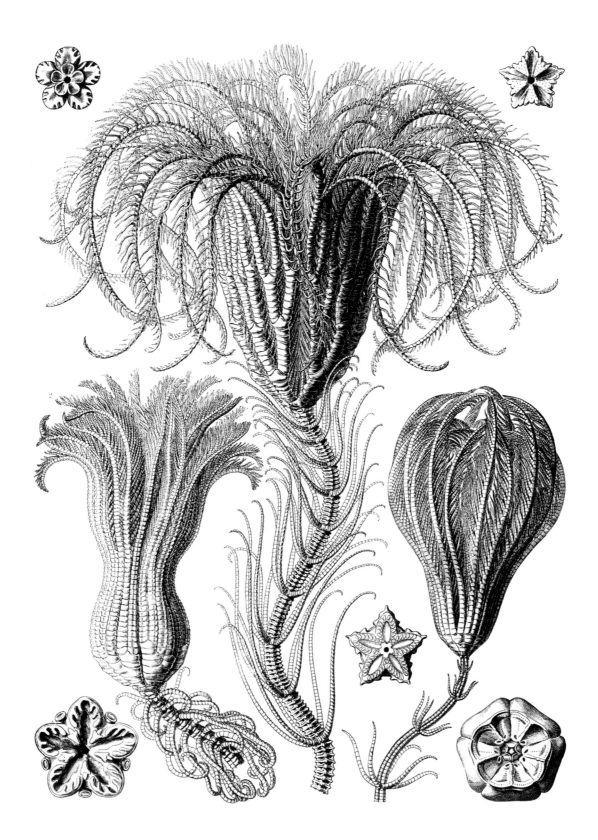

Opposite Crinoids can look like undersea feather dusters, but are, in fact, animals related to sea stars.

Above Crinoids are often anchored to the sea bottom by a stalk which ends in a crown of arms surrounding the mouth and body.

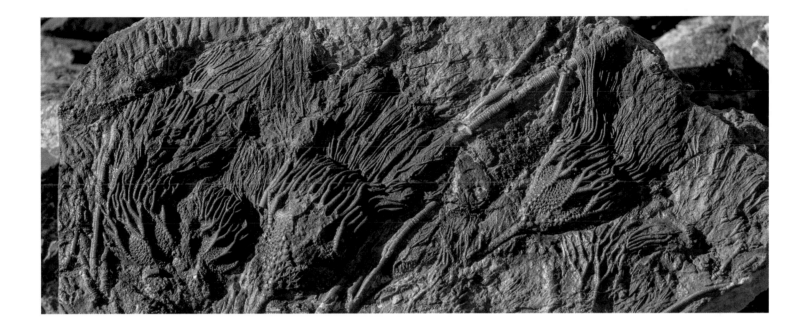

species – not too bad for animals that originated before any plants or animals lived on land.

Tracing the origins of crinoids is a challenging task. It's possible that the first crinoids date back to the time of the Burgess Shale and animals like *Anomalocaris*, over 508 million years ago. A strange fossil called *Echmatocrinus* might be the earliest crinoid, although it could also be a form of coral. The earliest definitive crinoids, then, date to rocks about 480 million years old from a time called the Ordovician. And from that point, crinoids underwent an impressive evolutionary explosion. Dozens and dozens of species evolved to filter food from the water, and as crinoid predators evolved – such as the earliest fish – these sea lilies and feather stars began evolving in new species. Some crawled and swam rather than staying in one place. Others became incredibly spiny and almost club-like in appearance. And despite almost going extinct during a terrible mass extinction 252 million years ago, some crinoids persisted and began to proliferate anew during the Age of Dinosaurs and beyond. Perhaps they might not be as charismatic as a giant squid or anglerfish, but crinoids certainly have staying power.

Crinoids can be found at almost any depth, from near the surface to approximately 9,000 metres (29,527 feet) down. This is a relatively recent find. Even though oceanographers have trawled up evidence of deep-sea crinoids before from depths below 6,000 metres (19,685 feet), it's only been recently that researchers have been able to visit those at the deepest depths. In 2009, a team of Japanese researchers spotted stalked crinoids over 9,000 metres down in the Izu-Ogasawara Trench, off the island nation's eastern coast. The crinoids were small, with arms only about 10 centimetres (4 inches) across, anchored to hard, rocky parts of the sea bottom. Experts did not recognize the species, made more challenging by the fact that they could not collect the specimen, but the find left no doubt that crinoids have been able to make their home in the deep with just as much ease as the upper levels of the ocean.

While many organisms in the deepest zone of the oceans seem relatively rare or specifically adapted to the harsh conditions, the crinoids caught on video by the Japanese scientists didn't seem all that different from others. The researchers noted that the deep crinoids were abundant and showed the same feeding postures as others. While it's possible that there's an unknown hydrothermal vent nearby, the researchers found no evidence of such a deep-sea oasis. Instead, it's more likely that the Izu-Ogasawara Trench is a kind of natural trap for organic matter from higher ocean levels – enough to let crinoids hold tight at extreme depth.

Opposite Illustration of crinoids from the Devonian period, which floated upside down in the water column.

Above Crinoids were more numerous and diverse in the past, with today's species representing only a fraction of what once existed.

Following pages Crinoids attached to wreckage off the Solomon Islands.

Trieste

IT'S ONE THING TO KNOW THAT THE OCEAN IS INCREDIBLY DEEP. IT'S ANOTHER TO VISIT THOSE PLACES. EVER SINCE THE DISCOVERY OF THE CHALLENGER DEEP, OCEANOGRAPHERS HAVE WONDERED ABOUT WHAT MIGHT BE DOWN AT SUCH INCREDIBLE DEPTH. IN 1960, A SMALL CREW WOULD GET THE FIRST DIRECT LOOK AT THIS DISTANT PLACE IN THE SUBMERSIBLE *TRIESTE*.

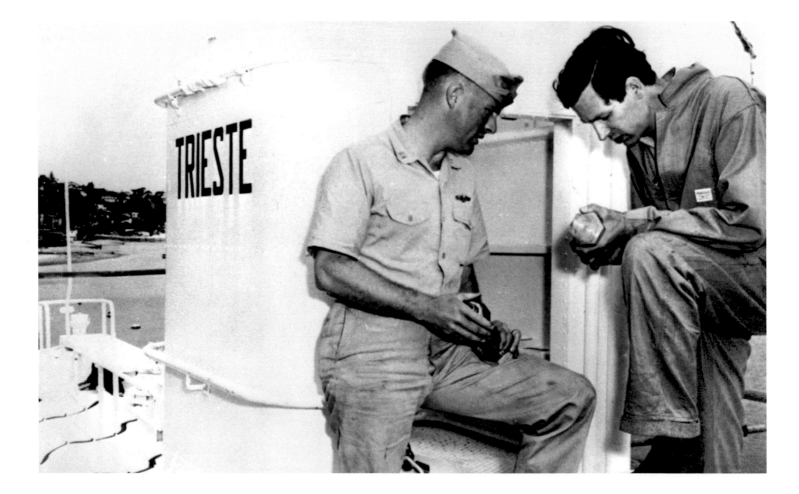

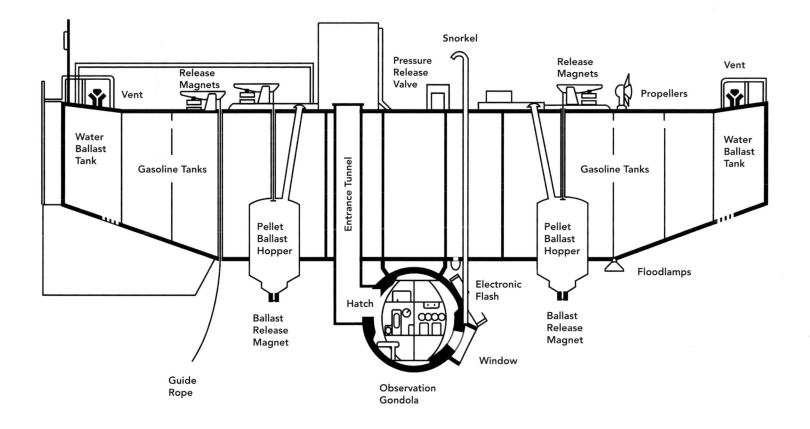

Opposite Don Walsh and Jacques Piccard aboard *Trieste* in 1960.

Above General arrangement of *Trieste*. The bathyscaphe was operated by the French Navy before being purchased by the US Navy in 1958.

Designed by Swiss engineers and built in Italy, *Trieste* was something of a halfway point between the bathysphere used by William Beebe and Otis Barton (see pages 80–83) and more modern deep-sea vehicles like the ever-updated *Alvin* (see pages 168–173). Despite being over 18 metres (59 feet) in length, most of *Trieste*'s bulk was made up of chambers to hold gasoline and regulate pressure. Crew space wasn't much more spacious than Beebe and Barton had in their bathyscaphe. The *Trieste* had a crew of two, held within an observation gondola suspended below the vessel. The explorers would enter from the top of *Trieste*, going through the midsection of the vessel to reach the hatch into their windowed bubble. But the *Trieste* was designed to do something entirely different from the comparatively shallow dives of preceding submersibles. There were no air hoses or cables connected to the *Trieste*. The vessel was designed to make a free dive to the bottom, sinking and rising on its own.

The *Trieste* used a combination of technologies to move up and down through the water column. Naturally, how to gracefully sink down to extreme depth and return was a huge consideration. To go down, crews loaded the *Trieste* with 9 metric tonnes (19,841 pounds) of iron pellets in interior silos. These were held in by powerful magnets, allowing the ballast to be released at depth as

well as in the case of electrical problems. The ship's buoyancy was regulated by gasoline. That might seem like a strange choice, but it came down to physics. Gasoline is less dense than water and does not compress as easily under great pressure, assuring an ascent if the *Trieste* ran into trouble at depth. The engineers behind the design did not want to take chances. Even the sphere attached to the bottom of the *Trieste* was designed to be 12.7 centimetres (5 inches) thick and able to withstand greater pressures than were expected. No one had been down so deep before, and estimating the conditions the crew might find was more difficult than planning a launch outside the atmosphere.

Originally operated by the French Navy, the *Trieste* was sold to the United States Navy in 1958. The American military had big plans for the submersible. Under Project Nekton, the US Navy planned several deep dives in the Pacific with the goal of taking measurements related to its performance – from effects of water pressure on the hull and how far light reaches to where life in the sea might be found. Following a few updates to the original

design, the *Trieste* was carried out to the waters off Guam in 1959 with the ultimate goal of exploring the Challenger Deep.

Following surveys that used depth charges, rather than weighted rope, to find the bottom, the crew of the *Trieste* started their deep dive on 23 January 1960. Aboard was Jacques Piccard – son of Auguste Piccard, who designed the *Trieste* – and Don Walsh of the US Navy. Dropping at nearly 1 metre (3 feet) per second, the descent to the bottom took 4 hours and 48 minutes. It was a harrowing trip. At about 9,000 metres (29,527 feet), one of the outer Plexiglass window panes of the observation bubble cracked. But Piccard and Walsh continued. Eventually they reached a depth of 10,911 metres (35,797 feet). It was so far beneath the surface than it took about 7 seconds for voice messages from the *Trieste* to

reach the surface while the crew inside shivered in the 7°C (47°F) cold and ate chocolate bars. They only spent 20 minutes on the bottom before their ascent, but that was long enough to know that even the deepest parts of the ocean supported life. Over a mucky ocean bottom made of decomposing diatoms, Piccard and Walsh saw flounder swimming. Even though humans, wrapped in tons of technology, could just barely visit the Challenger Deep, life had found a way to adapt to one of the harshest environments on the planet.

Above Final checks are made to the *Trieste* before its record dive to the Mariana Trench.

Opposite The crew was confined to the observation gondola suspended below the main vessel.

Mariana Trench

THE MARIANA TRENCH IS EARTH'S DEEPEST PLACE. IN THE WHOLE OF
THE WORLD'S OCEANS, EVEN AMONG THE ABYSSAL ZONE HABITATS
WHERE ORGANISMS MUST EKE OUT A LIVING IN THE PERPETUAL
AQUATIC DARK, THERE IS NO PLACE THAT COMPARES.

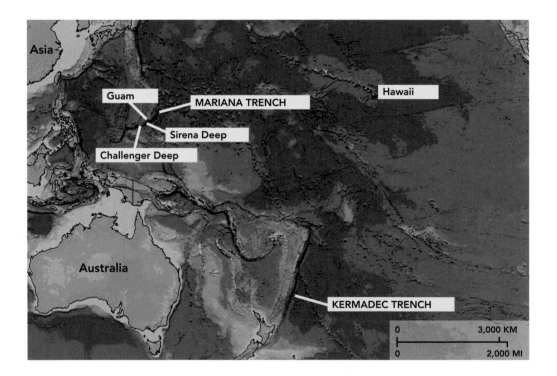

The Mariana Trench is not a single, even point along the ocean bottom. Instead, this grand crevice stretches about 2,500 kilometres (1,553 miles) in a crescent-shaped slice in the western part of the Pacific Ocean. The deepest part of the trench – the portion that has transfixed human imaginations for decades – is 10,984 metres (36,036 feet) below the surface, a dark place experts call the Challenger Deep. The incredible pressure in these depths reaches over 1,000 times the pressure at sea level, equivalent to about 50 jumbo jets piled on top of one another. Only a few people, in specially designed craft, have ever made brief visits here.

The Mariana Trench is geologically complex and was created by the motion of Earth's crust. The process works like this: the rocky outer shell of our planet is made up of about seven massive plates, and these plates move. As magma from within the Earth oozes out of ocean ridges and creates new crust, the older and heavier parts of the crust begin to sink and get pushed beneath other plates. Additional forces can be at play – like heat from the Earth's mantle – but the manner in which new crust is formed and is subsumed is thought to be the primary way these plates get pushed around. All this moving and shaking helps to create Earth's geological profile, including the Mariana Trench.

Rather than being a standalone landmark, the Mariana Trench is part of an ocean subduction zone. This is a place where heavier, thicker, older rock is pushed below a plate. The Pacific Plate that underlies much of the Pacific Ocean is relatively old, with rock that formed during the middle of the Jurassic period, and

that dense, heavy body of rock is slowly being shoved beneath the comparatively lighter Mariana Plate to the west. What the Mariana Trench really is, then, is the deepest part of the boundary between these two great geologic provinces.

But the fact that this great subduction zone remains so deep has to do with another quirk of location. Rivers and other moving bodies of water on a continent often drain to the sea, carrying tons upon tons of sediment with them. Divots on the ocean floor can become filled in over time if they're close enough to a coastline. But while many of these trenches and subduction zones

Opposite Map of the Mariana Trench, showing the location of Challenger Deep.

Below Parts of the *Trieste* were overbuilt to make sure the vessel could withstand pressures that no submersible had ever experienced before.

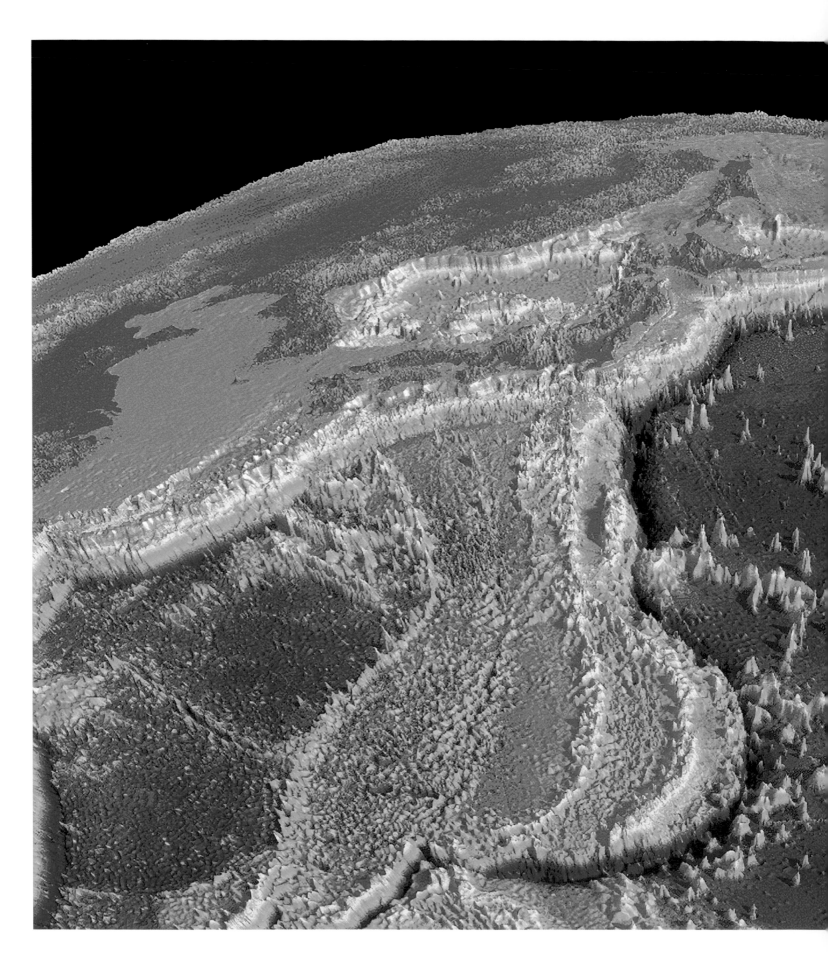

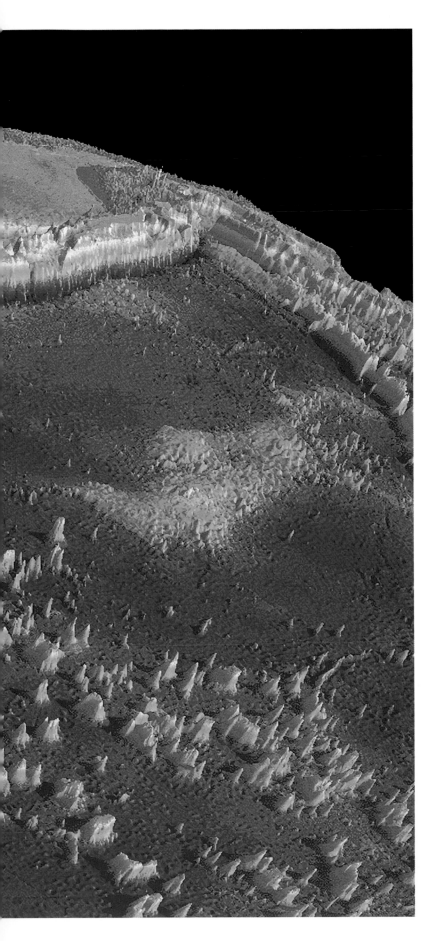

are close to continents, the Mariana Trench is far from land and therefore has not been infilled over time.

Scientists and naturalists have been fascinated with the Mariana Trench since HMS *Challenger* took the first sounding of the depths in 1875 as part of the ship's three-year oceanographic survey (see pages 196–201). Even then, no one fully understood just how far down the trench went – the initial reading was 8,183 metres (26,847 feet). Later expeditions that used a variety of techniques – from echo soundings and sonar to rare visits with submersibles – found even greater depths.

Despite the legendary status of the Mariana Trench as a realm of darkness and monstrous creatures in fiction, life has actually been very difficult to find in the deepest part of the oceans. In 2017, for example, researchers were thrilled to spot a pale snailfish at a depth of 8,178 metres (26,830 feet) in the trench. (Some previous fish sightings, such as a flounder supposedly seen in 1960, have been questioned or revised over time.) In 2014, another crew found a strange form of life – shrimp-like crustaceans called amphipods that grew far larger than their counterparts in the upper levels of the ocean. While many amphipods are about 2 to 3 centimetres (¾–1¼ inches) long, these "supergiants" reached up to 10 times that length.

The amphipods are not the only giants in the depths. At 10,641 metres (34,911 feet) down, a 2011 expedition into the trench spotted immense, strange creatures that may be the largest individual cells on the planet. Technically called xenophyophores, these organisms are like soft-bodied amoebas with a hardened outer shell. While so common in the oceans that they are often little studied, some of the xenophyophores in the Mariana Trench were over 10 centimetres (4 inches) across – exceptionally large for a single-celled organism.

Why these strange creatures live at such great depths is still unknown. But finding bigger versions of familiar organisms isn't an unexpected phenomenon. Zoologists know this as deep-sea gigantism, where organisms like crustaceans, squid and eels tend to be larger with increasing depth. (See also sea spiders, pages 182–185 and giant isopods, pages 192–195.) Even though life in the deepest part of the seas is strange and relatively rare, it has undoubtedly found a way.

Left Computer model of the topography around the Mariana Trench (indicated by the purple arc) using data from ship soundings and satellite altimetry.

Glossary

ALGAE
Photosynthetic organisms belonging to several different lineages, ranging from single cells to kelp.

BATHYSCAPHE
A free-diving vessel that can move under its own propulsion for deep-sea exploration.

BIOGENIC OOZE
Sediment made up of decomposing microorganisms and other organic matter, often plankton such as diatoms and foraminiferans.

CAMBRIAN PERIOD
The time period extending from 541 to 485.5 million years before the present day, a time during which early animal life proliferated.

CEPHALOPOD
A group of invertebrate animals, such as squid, octopus, cuttlefish and nautilus, that have bilateral body symmetry and a head ending in arms or tentacles.

CETACEAN
Fully aquatic mammals, including toothed and baleen whales, descended from land-dwelling ancestors.

CHEMOSYNTHESIS
A biological process that converts molecules containing carbon into a source of energy.

CONTINENTAL SHELF
The portion of a continent submerged beneath the ocean, usually shallower than the continental slope that leads to the Abyssal Plain.

CONVERGENT EVOLUTION
When two different species or broader categories of organisms independently evolve similar shapes, behaviours, or genetics, such as the comparable streamlined shapes of sharks and ichthyosaurs.

COUNTERSHADING
A coloration pattern that is darker above and lighter below and acts as camouflage.

CRETACEOUS PERIOD
The timespan between 145 and 66 million years ago, or the third period in the Mesozoic era, after the Triassic and Jurassic.

CRUSTACEAN
A diverse group of arthropods that include crabs, shrimp, krill, isopods, copepods and others.

DEPTH SOUNDING
Measuring the depth of a body of water, which can be done by mechanical means with a weighted tether or technologies such as sonar.

DIEL VERTICAL MIGRATION
The daily movement of sea organisms, especially plankton and the fish that feed on them, up towards the surface at night and to lower depths during the day.

DREDGE
An apparatus that brings samples up from depth, such as sediment or undersea organisms.

ELECTRORECEPTION
The natural ability to detect weak electrical fields, as in the jelly-filled pores of sharks.

FILTER-FEEDING
The behaviour of straining small organisms or organic matter from the water column by means of baleen, gills or some other apparatus.

FOSSIL

A trace of ancient life more than 10,000 years old, including body fossils such as bones and trace fossils such as burrows.

HYDROTHERMAL VENT

A fissure or opening in the seafloor from which heated, mineral-rich water rises.

ICHTHYOSAUR

Marine reptiles that evolved shark-like body shapes and lived between 250 and 90 million years ago.

JURASSIC PERIOD

The timespan between 201 and 145 million years ago, or the second period of the Mesozoic era, between the Triassic and Cretaceous.

MIDWATER

The part of the oceans between the surface and the bottom, where organisms live suspended in the water.

MOLLUSC

A phylum of invertebrates including clams, snails and squid, all of which have an organ called a mantle used for breathing and excretion.

NOTOCHORD

A flexible rod in the bodies of some animals made up of a cartilage-like material.

PALAEOZOIC ERA

The timespan between 541 and 251 million years ago, preceding the Mesozoic era.

PHOTOPHORE

A gland-like organ on an organism that produces light.

PHOTOSYNTHESIS

The natural process by which plants and other organisms use sunlight to make food from carbon dioxide and water.

PHYTOPLANKTON

Ocean-going microorganisms capable of photosynthesis.

ROV

Remote operated vehicle, or an undersea exploration vehicle operated by crew in another vessel.

SARCOPTERYGIAN

A group of bony fish with lobe-shaped fins, including the living coelacanth species.

SCLERAL RING

A circular arrangement of thin, plate-like bones in the eyes of some vertebrates, such as fish and marine reptiles.

SONAR

Sound navigation and ranging, or a technique that uses sound to communicate, detect and find objects or navigate.

STROMATOLITE

Dome-shaped geologic structure created by a mat of photosynthetic cyanobacteria adhering sediments together beneath it.

SUBMERSIBLE

Any vehicle capable of movement underwater, including remote operated vehicles and submarines.

TENTACLE

A grasping appendage that's flexible and elongated. In squid, there is one pair of tentacles in addition to eight arms.

TEST

A shell, often used to describe the silica-based shells of diatoms in comparison to shells made of calcium carbonate.

TRENCH

Long, narrow canyons, valleys or depressions in the seafloor reaching kilometres below the general surface of the sea bottom.

TRIASSIC PERIOD

The timespan between 251 and 201 million years ago, the first period of the Mesozoic era, before the Jurassic and Cretaceous.

ZOOPLANKTON

Tiny planktonic animals, ranging from diminutive crustaceans to the larvae of larger animals.

Index

Numbers in **bold** refer to main subjects, including
photos; in *italic* to all other photos/captions.

Picture Credits

The publishers would like to thank the following sources for their kind permission to reproduce the pictures in this book.

6-7 Dr Norbert Lange/Shutterstock, 9 Irina Markova/Shutterstock, 10 Juergen Faelchle/Shutterstock, 11 University of Rhode Island, 12 Littlesam/Shutterstock, 13 Mevans/Getty Images (top), 13 VectorMine/Shutterstock (bottom), 14 Humberto Ramirez/Getty Images, 15 Wei Huang/Shutterstock, 17 Oliver Denker/Shutterstock, 18-19 NASA, 20 Julie Ann Quarry/Alamy Stock Photo, 21 J. Roman/J.McCarthy, 22 Seb C'est Bien/Shutterstock, 23 Pictorial Press Ltd/Alamy Stock Photo, 24-25 Choksawatdikorn/Shutterstock, 27 Dante Fenolio/Science Photo Library (top), Mark Macewan/Nature Picture Library/Science Photo Library, 28 Helmut Corneli/Alamy Stock Photo, 29 Nick Hobgood, 30-31 Dante Fenolio/Science Photo Library, 32 FLHC A22/Alamy Stock Photo, 33 Awashima Marine Park/Getty Images, 34 Nature Picture Library/Alamy Stock Photo, 35 Awashima Marine Park/Getty Images, 36 James Ratchford/Shutterstock, 37 Blickwinkel/Alamy Stock Photo, 38 RLS Photo/Shuttestock (top), Science Photo Library/Alamy Stock Photo (bottom), 39 Dr Norbert Lange/Shutterstock, 40 Leighton Taylor/Department of Land Natural Resources, 41 Marko Steffensen/Alamy Stock Photo, 42-43 Biosphoto/Alamy Stock Photo, 44 Public Domain, 45 Bluehand/Shutterstock (top), Nature Picture Library/Shutterstock (bottom), 46-47 Gerard Lacz/Shutterstock, 48-49 Azoor Photo/Alamy Stock Photo, 51 Well/BOT/Alamy Stock Photo (left), The Natural History Museum/Alamy Stock Photo (right), 53 NASA/Alamy Stock Photo, 54 Alan Sirulnikoff/Science Photo Library, 55-57 Dotted Yeti/Shutterstock, 58 Public Domain, 59 Dante Feolio/Science Photo Library, 60-61 Leonid Serebrennikov/Alamy Stock Photo, 62 Yuri Photolife/Shutterstock, 63 Danny Ye/Shutterstock/Shutterstock (top), World History Archive/Alamy Stock Photo (bottom), 64 The Natural History Museum/Alamy Stock Photo, 65 Universal History Archive/Universal Images Group via Getty Images, 66 Zip Lexing/Alamy Stock Photo, 67 Rawpixel, 68 Adisha Pramod/Alamy Stock Photo, 69 Ewald Rübsamen via Wikimedia Commons, 71 cbpix/Alamy Stock Photo, 72 Reinhard Dirscherl/Alamy Stock Photo, 73 Nick Veasey/Science Photo Library, 75 Robert Harding/Alamy Stock Photo, 76 Fossil & Rock Stock Photos/Alamy Stock Photo (top), Corbin17/Alamy Stock Photo (bottom), 77 Mark Garlick/Science Photo Library, 78-79 Tagliaferri Photography/Alamy Stock Photo, 81 Bettmann/Getty Images, 82 Sueddeutsche Zeitung Photo/Alamy Stock Photo, 83 Granger Historical Picture Archive/Alamy Stock Photo, 84 Corbis/Getty Images, 85 Martin Almqvist/Alamy Stock Photo, 86-87 Chokswatdikorn/Shutterstock, 89 Marko Steffensen/Alamy Stock Photo (top), Alamy Stock Photo (bottom left), Granger Historical Picture Archive/Alamy Stock Photo (bottom right), 90-91 Kelvin Aitken/VWPics/Alamy Stock Photo, 93 Classic Images/Alamy Stock Photo, 94 David McNew/Getty Images, 95 Calimax/Alamy Stock Photo (top), Reuters/Alamy Stock Photo (bottom), 96-97 By Wildestanimal/Getty Images, 99 Blue Planet Archive/Alamy Stock Photo (top left), Doug Perrine/Alamy Stock Photo (bottom left), Personnel of NOAA Ship PISCES (right), 101 Dotted Zebra/Alamy Stock Photo (top), VTR/Alamy Stock Photo (bottom), 102 Eric Broder Van Dyke/Alamy Stock Photo, 103 Minden Pictures/Alamy Stock Photo, 105 Nature Picture Library/Alamy Stock Photo, 106 Mauritius Images Gmbh/Alamy Stock Photo, 107 Morgan Trimble/Alamy Stock Photo, 109 NOAA, 110 Dante Fenolio/Science Photo Library, 111 Mauritius Images Gmbh/Alamy Stock Photo, 112-113 Dante Fenolio/Science Photo Library, 115 NOAAA/Craig Smith (top), Subphoto.com/Shutterstock (bottom), 116 Adisha Pramod/Alamy Stock Photo (top), Mark Conlin/Alamy Stock Photo

(bottom), 117 The Natural History Museum/Alamy Stock Photo, 118 Frank Hecker/Alamy Stock Photo, 119 Nature Photographers Ltd/Alamy Stock Photo, 120-121 Gina Kelly/Alamy Stock Photo, 122 Julian Partridge, 123 Universal Images Group North America LLC/Alamy Stock Photo (top), Minden Pictures/Alamy Stock Photo (bottom), 124-125 Minden Pictures/Alamy Stock Photo, 126 Richard Ellis/Science Photo Library, Diarmuid/Alamy Stock Photo (bottom), 129 Didier Descouens, 130 Sabena Jane Blackbird/Alamy Stock Photo, 131 Photograph taken by Mark A. Wilson (Department of Geology, The College of Wooster), 133-134 Dante Fenolio/Science Photo Library, 135 Neil Bromhall/Shutterstock, 136 Entertainment Pictures/Alamy Stock Photo, 137 EyeEm/Alamy Stock Photo, 138 Science History Images/Alamy Stock Photo, 141 Adisha Pramod/Alamy Stock Photo, 143 WHOI (top), Caryolyn Ruppel/Woods Hole Coastal and Marine Science Center/US Geological Survey/Science Photo Library (bottom), 145 Blue Planet Archive/Alamy Stock Photo, 146-147 NOAA Okeanos Explorer Program, Galapagos Rift Expedition 2011/Science Photo Library, 148 Nature Picture Library/Alamy Stock Photo, 149 Fran Martin de la Sierra/Alamy Stock Photo, 150 Minden Pictures/Alamy Stock Photo, 151 Mark Conlin/Alamy Stock Photo, 152 David Herraez Calzada/Alamy Stock Photo, 153 Craig Lambert Photography/Alamy Stock Photo, 154 Tryton2011/Shutterstock, 155 Wirestock Creators/Shutterstock, 156 Public Domain, 157 OpenCage Creative Commons Attribution-Share Alike (top), Andrea Izzotti/Shutterstock, 158 Library Book Collection/Alamy Stock Photo, 159 Andrea Izzotti/Alamy Stock Photo, 160 Hectonichus via Wikimedia Commons, 161 Geography Photos/Universal Images Group, 162-163 Falconaumanni via Wikimedia Commons, 165 Blickwinkel/Alamy Stock Photo (top), DeAgostini/Getty Images (bottom), 166 Universal History Archive/Universal Images Group (top), 167 Scenics & Science/Alamy Stock Photo, 168 Getty Images/Bettmann, 169 Science History Images/Alamy Stock Photo, 170 Woods Hole Oceanographic Institution, 172-173 Kirt L. Onthank, Creative Commons Attribution-Share Alike, 175 ThreeArt/Alamy Stock Photo, 176 Silke Baron, 177 Nhobgood via Wikimedia Commons, 178 Catmando/Shutterstock, 179 The Five Deeps Expedition, 180-181 Nature Picture Library/Alamy Stock Photo, 182 David Chapman/Alamy Stock Photo, 183 MNHN, 184 Simon Brockington/Shutterstock, 185 Cbimages/Alamy Stock Photo, 187 Adisha Pramod/Alamy Stock Photo, 188 David Shale/Nature Picture Library, 189 Image courtesy of Journey into Midnight - Light and Life Below the Twilight, 190-191 Nature Picture Library/Alamy Stock Photo, 193 Kikujungboy CC/Shutterstock, 194 Tony Wu/Nature Picture Library, 195 Ted Kinsman/Science Photo Library (top), Tony Wu/Nature Picture Library (bottom), 196 History and Art Collection/Alamy Stock Photo, 197 Mansell Collection/The LIFE Picture Collection/Shutterstock, 198-199 NHM Images, 200 Science & Society Picture Library/SSPL/Getty Images, 202 Alexander Vasenin, 203 Ernst Haeckel, 204 Richard Bizley/Science Photo Library, 205 Phil Degginger/Science Photo Library, 206-207 Stocktrek Images, Inc/Alamy Stock Photos, 208 Everett Collection Inc/Alamy Stock Photo, 209 Ralph Sutherland via Wikimedia Commons, 210-211 US Navy/Science Photo Library, 212 Science History Images/Alamy Stock Photo, 213 U.S. Naval History and Heritage Command, 214-215 US Geological Survey/Science Photo Library

Every effort has been made to acknowledge correctly and contact the source and/or copyright holder of each picture and Welbeck Publishing Group apologises for any unintentional errors or omissions, which will be corrected in future editions of this book.

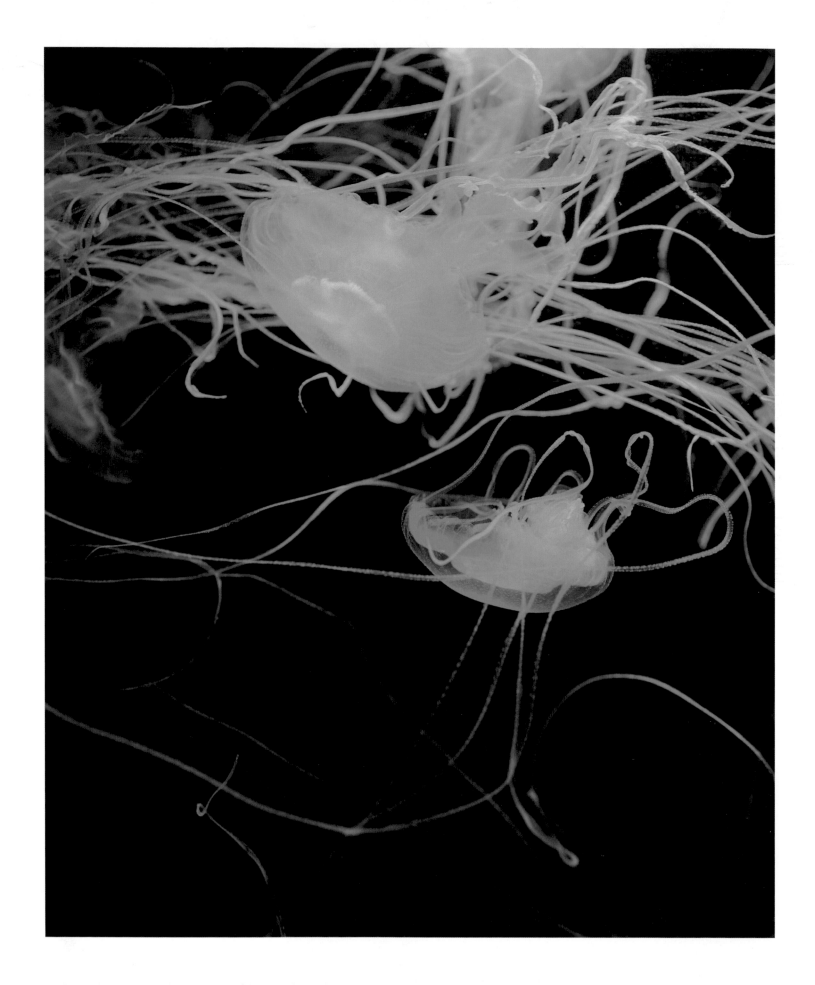

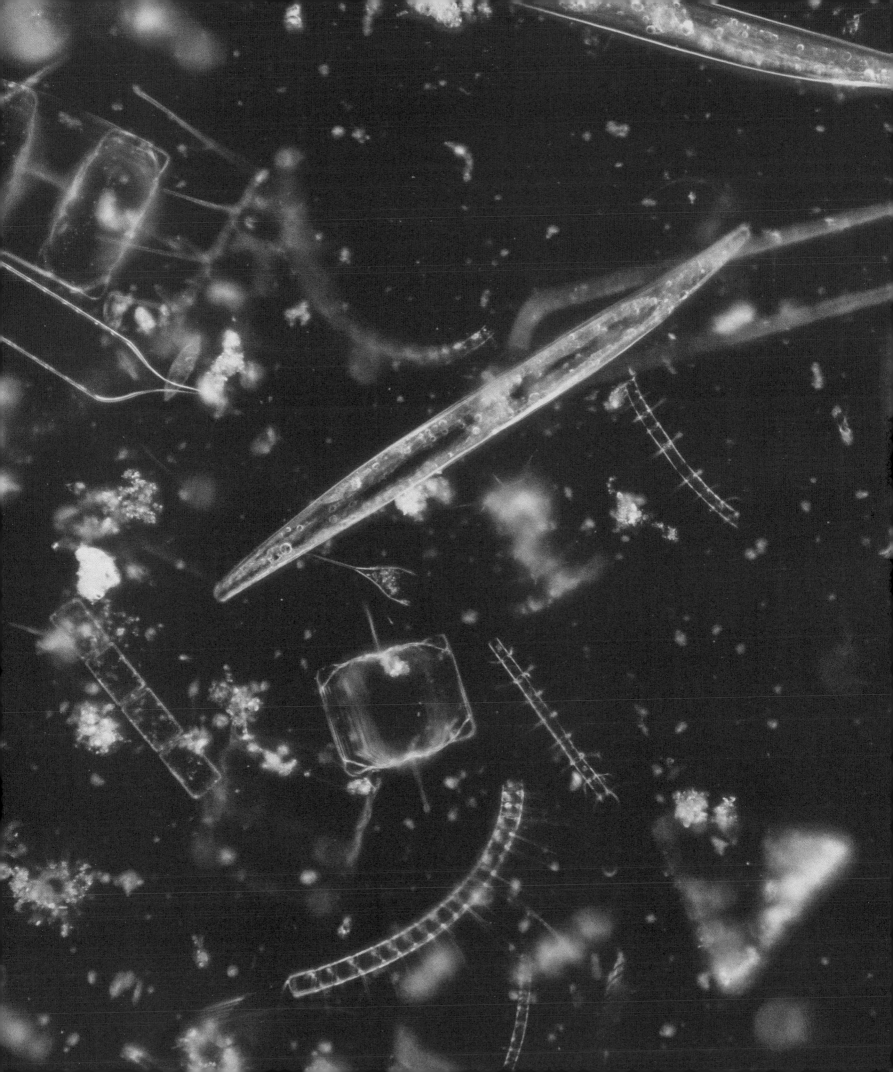

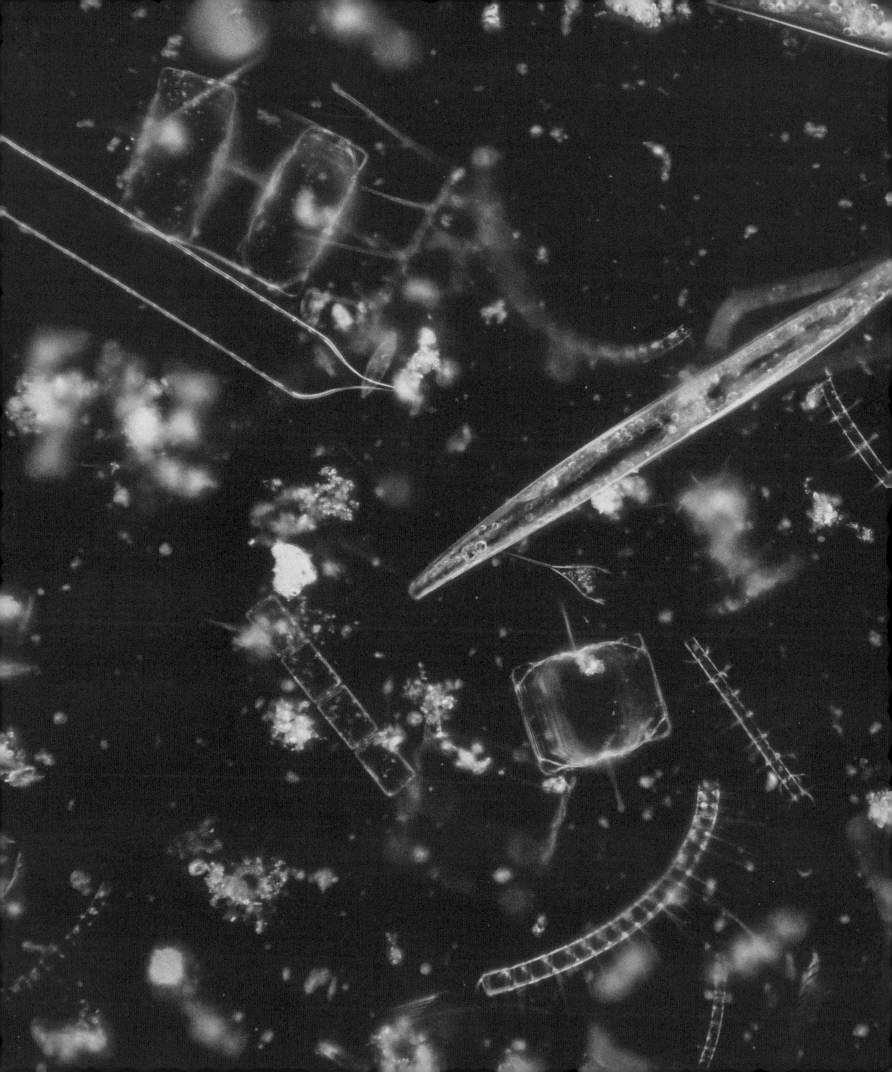